Biopesticides in Horticultural Crops

This book is a compendium of information related to innovations, commercialization and registration of biopesticides, recent advances in mass production, formulation, extension of shelf life, delivery systems of antagonists and entomopathogens and synergistic and antagonistic response of biopesticides with agrochemicals. The information on all the important laboratory protocols and techniques in isolation, identification, selection, culturing, mass production, formulation, enhancement of shelf life and biosafety issues of bioinoculants used as Biopesticides in Horticulture crops such as Antagonistic Fungi and Bacteria, Actinomycetes, Entomopathogenic Fungi, Bacillus thuringiensis, Nuclear Polyhedrosis Viruses and Entomopathogenic Nematodes have been included for the benefit of research scientists, teachers, research scholars and students working in the field of Biopesticides.

Dr. Rajeshwari R. is presently working as Assistant Professor in the Department of Plant Pathology at College of Horticulture, Mysuru. She is the recipient of University Gold Medal and Late Dr. K. Ramakrishna (Ex-Dean, UAS) Memorial Gold Medal for PhD in Plant Pathology from UAS, GKVK, Bengaluru for best virology thesis of the year 2014. She was awarded Travel grant and research fellowship from European Molecular Biology Organisation (EMBO) and Centre for International Co-operation in Science to undergo training at European Molecular Biology Laboratory (EMBL), Heidelberg, Germany during 2014.

Dr. Vikram Appanna secured his Bachelors' degree in Agriculture from University of Agricultural Sciences, Bangalore, Master's and Ph.D. degrees in Agricultural Microbiology from University of Agricultural Sciences, Dharwad (1991-2001). He worked as a Post Doctoral Fellow in Plant Science Department of McGill University, Quebec, Canada from 2002 to 2006. He also worked as a Research Associate from 2006 to 2009 in Potato Development Centre, New Brunswick Department of Agriculture and Aquaculture, Wicklow, New Brunswick, Canada

Biopesticides in Horticultural Crops

Rajeshwari R.
Assistant Professor (Plant Pathology)
College of Horticulture, Bengaluru
University of Horticultural Sciences
Bagalkot-587104

Vikram Appanna
Assistant Professor (Agricultural Microbiology)
College of Horticulture, Mysuru
University of Horticultural Sciences
Bagalkot-587104

CRC Press
Taylor & Francis Group
Boca Raton London New York

CRC Press is an imprint of the
Taylor & Francis Group, an **informa** business

NARENDRA PUBLISHING HOUSE

First published 2022
by CRC Press
2 Park Square, Milton Park, Abingdon, Oxon, OX14 4RN

and by CRC Press
6000 Broken Sound Parkway NW, Suite 300, Boca Raton, FL 33487-2742

British Library Cataloguing-in-Publication Data
A catalogue record for this book is available from the British Library

Library of Congress Cataloging-in-Publication Data
A catalog record has been requested

ISBN: 978-1-032-15277-6 (hbk)
ISBN: 978-1-003-24342-7 (ebk)

DOI: 10.1201/9781003243427

Contributors

Dr. M.S. Rao	ICAR-Indian Institute of Horticulture Research Bangalore, Karnataka
Dr. G. Manjunath	Department of Plant Pathology College of Horticulture, Mysuru University of Horticultural Sciences Bagalkot, Karnataka
Dr. N. Geetha	Department of Biotechnology University of Mysore, Mysuru, Karnataka
Dr. Raghavendra M.P.	Postgraduate Department of Microbiology Maharani's Science College for Women Mysuru Karnataka
Dr. S. Sriram	Division of Plant Pathology ICAR-Indian Institute of Horticulture Research Bangalore, Karnataka
Dr. Rajeshwari R.	Department of Plant Pathology College of Horticulture, Bengaluru University of Horticultural Sciences Bagalkot, Karnataka
Dr. Ashwitha K.	Centre for Cellular and Molecular Platforms NCBS-TIFR, GKVK Bengaluru, Karnataka
Dr. S.S. Hussaini	ICAR- National Bureau of Agriculture Insect Resources, Bengaluru, Karnataka
Dr. Jagadeesh Patil	Division of Molecular Entomology ICAR-National Bureau of Agriculture Insect Resources, Bengaluru, Karnataka
Dr. G. K. Rame Gowda	Regional Horticultural Research and Extension Centre, UHS Campus, GKVK Bengaluru, Karnataka

Dr. Prasad Kumar

Department of Entomology
College of Horticulture, Mysuru
University of Horticultural Sciences
Bagalkot, Karnataka

Dr. M. Mohan

Division of Molecular Entomology
ICAR-National Bureau of Agriculture Insect
Resources, Bengaluru, Karnataka

Dr. Mahesh S. Yandigeri

Division of Molecular Entomology
ICAR-National Bureau of Agriculture Insect
Resources, Bengaluru, Karnataka

Dr. Kiran Kumar K.C.

Assistant Professor and Head
Horticulture Research and Extension Center
Arsikere, Hassan, Karnataka

Dr. Raghavendra A

Division of Insect Biosystematics
ICAR-National Bureau of Agriculture Insect
Resources, Bengaluru, Karnataka

Dr. Vikram Appanna

Department of Natural Resource Management
College of Horticulture, Mysuru
University of Horticultural Sciences
Bagalkot, Karnataka

Contents

Preface

I am pleased to present this book related to biopesticides in horticultural crops. The chapters of this book are a compilation of lectures delivered by eminent scientists on biopesticides during a short course sponsored by Indian Council of Agriculture Research (ICAR), New Delhi on "Recent advances in development of Bioformulations" held at College of Horticulture, Mysuru, Karnataka from 3rd to 12th September, 2018.

There is overwhelming evidence that pesticides pose a potential threat to humans and other life forms in nature. Hence, the new challenge in the new millennium is an alternative approach ie., use of microbial biopesticides. It offers greener and safer alternatives for increasing soil fertility, crop productivity and achieving the goal of sustainability. Microbial biopesticides are robust than synthetic chemicals, comprising of bioinoculants for plant protection either singly or in consortium. Biocontrol agents interact with pathogens and through multifaceted approach, suppress pests and diseases, improve plant resistance and detoxify toxic substances in soil. Inert carrier materials in biopesticides increases the establishment of antagonistic microbes for longer duration in plant and rhizosphere and enhances the production of antibiotics, siderophores, hydrolysing enzymes, phytohormones and other volatile extracellular metabolites. They make up only a small fraction of the total deliveries to plants due to host specificity, difficulties of culturing microbes in large quantities and development of stable formulations with improved shelf life and field efficacy.

The knowledge about isolation, selection, culturing, mass production, formulation, registration, biosafety issues and delivery systems of biopesticides in different agro-climatic conditions is essential to all the researchers involved in production of high quality products in our country. The different chapters were penned and presented by eminent scientists from different institutions like ICAR-IIHR, Bangalore, ICAR-NBAIR, Bangalore, University of Agricultural Sciences, Bangalore, University of Mysore, Mysore and University of Horticultural Sciences, Bagalkot.

I would like to thank ICAR, New Delhi for sponsoring the training programme on "Recent advances in development of Bioformulations" and Dr. Janardhan, G.,

Dean, College of Horticulture, Mysuru for his immense support and help in conducting this programme. I am thankful to the Course Director of this training programme, Dr. Vikram Appanna, Assistant Professor (Agricultural Microbiology), COH, Mysuru for his constant support and encouragement and for providing me an opportunity to be a part of this training programme as Course Coordinator.

Editors

1

Innovations, Commercialization and Registration of Biopesticides

M. S. Rao

ICAR-Indian Institute of Horticulture Research, Bangalore

Globalization and intense pressure on land owing to the accelerating population has led to diversification of agriculture and increasing area under protected cultivation. However, in the present scenario, vagaries due to biotic stresses become inevitable due to intensified cropping systems. Among the biotic stresses, plant parasitic nematodes have assumed an alarming stature as hidden enemies of crop, directly as pathogens and indirectly predisposing the plants to pathogens. Annual losses due to plant parasitic nematodes are estimated at US $100 billion worldwide and $40 million in India.

Since the horticultural produce especially fruits and vegetables are consumed afresh, consumers expect residue-free produce both for internal and export markets. There is a great need for environmentally-friendly microbial technologies in agriculture. There is a need for restructuring the crop rhizospheres for improving and sustaining the nutrient supply in the soils and enhancing the health and yield of crops through sustainable practices based on technologies such as 'biopesticides'.

Biopesticides are certain types of pesticides derived from natural materials such as animals, plants, bacteria, and certain minerals. Microbial biopesticides include the principles of microbial ecology, which encompass inoculation of crops with beneficial microorganisms and the use of cultural practices that enrich indigenous beneficial microorganisms in individual agricultural fields. Biopesticides have overwhelming advantages of high selectivity to target pests, safety to biotic life and environment, amenability to individual applications, integrated pest management and suitability for organic niche products including export oriented commodities. These supports have reduced usage of chemical pesticides leading to eco-friendly environment and chemical-free produce for consumption. Management strategies based on biopesticides helps not only in controlling nematode diseases but also enhanced the growth of plants resulting in increased yield and enhancing socio-economic status of farmers.

The author of this chapter, Dr. M. S. Rao, Principal Scientist, Division of Entomology & Nematology, ICAR- Indian Institute of Horticultural Research, Bengaluru, India is the pioneering researcher who commercialized the biopesticides all over India. He has registered five biopesticide products, namely *Paecilomyces lilacinus* 1 % W.P., *Pochonia (Verticillium) chlamydosporia* 1 % W.P., *Trichoderma viride* 1.5 %, *Trichoderma harzianum* 1 % W.P., and *Pseudomonas fluorescens* 1 % W.P. with the Central Insecticides Bureau & Registration Committee (CIB & RC), New Delhi under section 9 (3b) of the Insecticides Act, 1968 for managing nematode problems and associated disease complexes in horticultural crops in the country. These technologies are transferred to around 540 industries/licensees all over India licensed for commercial production of biopesticides.

The following patents were granted to Dr. M. S. Rao for bio-pesticide formulation technologies:

1. US Patent No.: US 7,923,005 B2, Granted on 12th April, 2011 - A process for producing a biopesticide composition containing *T. harzianum* and *P. fluorescens*.

2. European Patent No.: EP 2154975 B1, Granted on 7th November, 2012 - A process for production of organic formulation of biopesticide, *P. fluorescens*.

3. Australian Patent No.: AU 2007216174 B2, Granted on 31st January, 2013 - A process for producing a biopesticide composition containing *T. harzianum* and *P. fluorescens*.

4. Australian Patent No.: AU 2007352140 B2, Granted on 21st February, 2013 - A process for producing a biopesticide composition containing *T. harzianum* and *P. fluorescens*.

5. Sri Lankan Patent No.: 15228 (PCT/IN2007/000401), Granted on 13th May, 2010 - A process for the production of organic formulation of biopesticide composition containing *P. fluorescens*.

6. Thailand Patent No.: 7621, Granted on 20th Nov., 2013- A process for producing a bio-pesticide composition containing *T. harzianum* and *P. fluorescens*.

7. Bangladesh Patent No. 1005199 - Granted on 3rd April, 2014 - A process for the production of organic formulation of biopesticide *P. fluorescens*.

8. Indian Patent No. 250779 (434/DEL/2006), Granted on 27th Jan, 2012- A process for producing a biopesticide composition containing *T. harzianum* and *P. fluorescens*.

Bio-pesticide Technologies Commercialized at ICAR- IIHR

S. No.	Name of the technology	No. of industries who received technology and licensed to produce
1	Bio-nematicide – *P. lilacinus* 1% W. P.	86
2	Bio-pesticide - *T. harzianum* – 1% W.P	102
3	Bio-pesticide - *P. fluorescens*– 1% W.P.	159
4	Bio-pesticide - *T. viride* – 1.5% W.P	164
5	Bio-nematicide - *V. chlamydosporium (P. chlamydosporia)* - 1% W.P	40
6	Arka plant growth enhancer and yield promoter (Patented technology)	1
7	Total number of industry licensees	552

Registration of Biopesticides

In India the use of pesticides is regulated by the CIB & RC, New Delhi. Any individual or organization intending to manufacture, import or export pesticides should register the product with CIB & RC. The CIB & RC then will anaylse the product for potential risk to human and animal health, its environmental impact and usefulness in agriculture. The use of pesticides is mainly governed by the Insecticides Act,1968, and Insectide Rule,1971. The CIB & RC has set guidelines for pesticides sale in the country. Any individual or company, intending to either manufacture or import any type of pesticides to India, needs to submit an application for the registration of their product to CIB & RC.

The submission of this application involves several technicalities. The application is to be submitted to the registration committee. The details regarding the active ingredients need to be specified. After receiving the application, along with the recommended fee, the committee conducts physical, chemical and microscopic tests and enquires to confirm the claims made by the manufacturer or importer as far as the utility of the pesticide is concerned. If the application is found to be valid, a Registration number is allocated and a Certificate of Registration is then made available within a period of twelve months.

Requirement of the data/information to be submitted by the subsequent applicant for getting registration under section 9(3) /9(3B) of already registered strain are as follows.

a. Form-I duly filled in along with requisite registration fee of Rs.100 as per existing requirement.

b. Already approved Label leaflets of the product/strain.

c. Testimonial/documents about the company as per existing requirement.

d. Undertaking about the strain from the inventor of the strain or first registrant or subsequent registrant of the strain or the applicant.

e. One sample (01 kg.) for Pre-Registration Verification (PRV) through Central Insecticides Laboratory as per already approved product specification by RC.

f. Another sample (01 kg) for PRV of Gene Code Sequencing of 16 Sr-DNA/finger printing along with a demand draft (as per invoice obtained as testing fee from NBAIM, Mau) in favour of NBAIM, Mau as testing fee for Gene code sequencing of 16 S r-DNA/finger printing.

g. Invoice for testing fee for Gene Code Sequencing of 16SR-DNA/finger printing has to be obtained from NBAIM, Mau by the Secretary of CIB & RC.

Registration of new strain of biopesticide

It was decided that the applicants for registration of new strain has to submit all the data as per existing guidelines for registration under section 9(3)/9(3B) for all the disciplines. Two samples have to be submitted to the Secretary of CIB & RC, one for Pre-Registration Verification (PRV) from Central Insecticides Laboratory as per product specification requirement & another sample to be used for pre-registration verification (PRV) of Gene code sequencing of 16 S r-DNA/finger printing along with a demand draft (as per invoice obtained as testing fee from NBAIM, Mau) in favour of NBAIM, Mau as testing fee for Gene code sequencing of 16 S r-DNA/ finger printing.

Minimum infrastructure required for production and registration of biopesticides

It was felt that flight-by-night manufacturers are available in the market and other products are sold in the name of biopesticides. Many of the manufacturers don't have proper and adequate infrastructure required for quality production of the biopesticides, as many samples of the biopesticides are failing in the test, which is affecting the quality of the produce. Therefore, verification of the infrastructure and technical competency of the applicants already registered under section 9(3B) and applying for registration u/s 9(3)/9(3B) extension has to be conducted by a team constituted by the Secretary, CIB & RC for the purpose.

Guidelines for Registration of Antagonistic Fungi under Section 9(3)/9(3b) of the Insecticide Act, 1968

Standard of Formulations

1. Colony Forming Unit (CFU) count on selective medium should be minimum of $2x10^6$ per ml or gm for *Trichoderma* spp.

2. Pathogenic contaminants such as gram negative bacteria, *Salmonella*, *Shigella*, *Vibrio* and such other microbials should not be present. Other microbial contaminants should not exceed $1x10^4$ count per ml or per gram of formulation. Chemical/botanical pesticide contaminants should not be present.

3. Stability of CFU counts at 30° C and 65% RH.

Registration requirement

S. No.	Requirements	9(3B)	9(3)
A	Biological characteristics and Chemistry for Registration		
1	a. Systematic name (Genus and species)	R	R
	b. Strain name	R	R
2	Common name, if any	R	R
3	Source of origin as Annexure -1.1	R	R
4	Habitat and morphological description	R	R
5	a) Composition of the product	R	R
	b) CFU/g of the product	R	R
	c) Percent content of the Biocontrol organism in the formulation & nature of biomass.	R	R
	d) Percentage of carrier/filler, wetting/dispending agent, stabilizers/emulsifiers, contaminants/impurities etc.	R	R
	e) Moisture content	R	R
6	Specification of the product as per Annexure-I	R	R
7	Manufacturing process including type of fermentation and biological end products. The microbial cultures are multiplied by liquid solid fermentation. Information pertaining to use of entire mycelial mats with spores separated must be provided in terms of biomass.	R	R
8	Test Methods:		
	a) Dual culture to attain at least 50% reduction in target organism.	R	R
	b) Bioassay: based on disease severity and root colonization as detailed in Appendix-I	R	R
9	Qualitative analysis	R	R
9.1	CFU on selective medium	R	R

9.2	Contaminants: a) Pathogenic contaminants such as *Salmonella, Shigella, Vibrio* and such other microbials	R	R
	b) Other microbial contaminants	R	R
	c) Chemical and botanical pesticide contaminants	R	R
9.3	Moisture content	R	R
9.4	Shelf life claims : Not less than 6 months	R	R
	a) Data on storage stability as per shelf life claims as detailed in Note-2	R	R
10	A sample for verification (100 g)	R	R
B	Bio-efficacy		
11	Field studies Data from SAU's/ICAR Institute certified by Director of Research of SAU or Head of the ICAR Institute	R*	R***
12	Laboratory studies: The product should be tested at a laboratory under ICAR/ SAU/CSIR/ICMR.	R	R
C	Toxicity		
13	For mother culture a) Single dose oral (rat and mouse)	R	R
	b) Single dose pulmonary	R	R
	c) Single dose dermal	R	R
	d) Single dose intraperitoneal	R	R
	e) Human safety records.	R	R
14	For formulation a) Data on mother culture as in (13) above	R	R
	b) Single dose oral (rat & mouse)	R	R
	c) Single dose pulmonary	R	R
	d) Primary skin irritation	R	R
	e) Primary eye irritation	R	R
	f) Human safety records	R	R

15	For formulated product to be directly manufactured (Mammalian toxicity testing of formulations)		
	a) Single dose oral (rat & mouse) Toxicity/Infectivity/Pathogenicity	R	R
	b) Single dose pulmonary Toxicity/Infectivity/Pathogenicity (Intratracheal preferred)	R	R
	c) Single dose dermal Infectivity	R	R
	d) Single dose intraperitoneal (Infectivity)	R	R
	e) Primary skin irritation	R	R
	f) Primary eye irritation	R	R
	g) Human safety records (effect/lack of effects)	R	R
16	Environmental safety testing: Core Information requirements (For formulation only)		
	a) Non-target vertebrates		
	Mammals[a]	NR	R
	Birds (two species)[b]	NR	R
	Fresh water fish[c]	NR	R
	b) Non-target invertebrates		
	Soil invertebrates[d]	NR	R
D	Packaging & Labeling		
17	Formulation Manufacturing process/process of formulation		
	a) Raw material	R	R
	b) Plant and Machinery	R	R
	c) Unit Process operation/unit process	R	R
	d) Out-put (finished product and generation of waste)	R	R
18	Packaging		
	a) Classification-solid, liquid or other types of product	R	R
	b) Unit pack size – In metric system	R	R
	c) Specification – Details of primary, secondary and transport pack	R	R
	d) Compatibility of primary pack with the product (Glass bottles are not recommended).	NR	R

19	Labels and leaflets As per Insecticides Rules, 1971/as per existing norms indicating the common name, composition, antidote, storage, statements etc	R	R

Abbreviations

R = Required

NR = Not Required

R** = Two seasons/year data on bio-effectiveness from minimum two agro-climatic conditions

R***= Two seasons/year data on bio-effectiveness from minimum three agro climatic conditions

a =Information on infection and pathogenicity in mammals will be available from mammalian safety testing

b =Information on infection and pathogenicity: suggested test: single-dose, oral test. suggested test species: pigeon and chicken.

c = Information on infection and pathogenicity: suggested test species: *Tilapia mossambica* or other appropriate spp.

d=Information on mortality effects. It is recommended that test species include an earthworm (***Lumbri custerrestris***) or other appropriate macro invertebrates of ecological significance.

1. Applicants are required to submit an undertaking that, strain is indigenous, naturally occurring, not exotic in origin, and not genetically modified as per Annexure 1.1.

2. Additional two months data for six months shelf-life claim, three months additional data for one year shelf-life claim at two different agro climatic locations at ambient temperature along with meteorological data should be submitted.

3. Considering the fact that many small entrepreneurs are engaged in the business of cultivation of antagonistic fungi the following simplification has been considered.

3.1 If same microbial strain is used for making formulation by different entrepreneurs, then the information submitted once on the said strain will be sufficient. All entrepreneurs need not to generate relevant data.

3.2 If same microbial strain, same method and same adjuvants, stabilizers etc. are used for making the given formulation, data once submitted for validating these claims will be sufficient for subsequent registrants, as substantiated by the relevant supportive documents.

4. The packaging material should also be ensured to be free from contamination from handling, storage and transportation and is as per prescribed standards, as the case may be.

Indian Standards for antagonistic fungi draft specifications

1. Form and appearance

2. pH

3. Composition

4. CFU/g of the product

5. Percent content of the biocontrol organism in the formulation and nature of biomass.

6. Percentage of carrier/filler, wetting/dispersing agent, stabilizers/emulsifiers, contaminants/impurities etc.

7. Moisture content

Method of Analysis

1. CFU counts by serial dilution and examination under regular compound research microscope with bright field optics.

2. Plating for contaminants on specific media.

3. Antagonistic mycolytic capability on target organism by bioassay on plants (laboratory test).

4. Bioassay procedure based on disease severity and root colonization.

5. An undertaking should be submitted that strain is indigenous, naturally occurring, not exotic in origin and not genetically modified as per Annexure-1.1.

2

Biorationals and Biopesticides for Disease Management of Horticultural Crops

G. Manjunath

University of Horticultural Sciences, Bagalkot

Introduction

The term "Biorational" implies for the products that are typically biologically-derived or, if synthetic, structurally similar and functionally identical to a biologically occurring material with minor differences between the respective stereochemical isomer ratios derived from biological or synthetic origins and has no environmental impact. "Biorational" include biopesticides as well as nonpesticide products but not limited to those used for crop stress management, enhanced plant physiology benefits, root growth management, postharvest treatments or as an alternative to pesticides (Muhammad Sarwar, 2015).

Overview of BCA Applications and Biorationals under Different Perspectives

Under changed climate many diseases turned to be epidemics in the recent times with high preferential management options. Whilst, other foliar and fruit rot diseases caused by species of *Alternaria, Colletotrichum, Cercospora* etc., also continue to have a serious impact on productivity. These diseases are chiefly managed by using synthetic pesticides of high cost and few of them have residual toxicity with possible chances of causing resistance in targeted pathogen, hence, the reduction in use of synthetic molecules in pomegranate is increasingly demanding. Further, most of synthetic molecules are counterproductive increasing inputs cost and unsafe to health and environment. Therefore, use of synthetic molecules have to be critically evaluated and it is favourable to replace /integrate chemical control methods with less toxic biological methods.

The use of bioagent formulations comprising *Pseudomonas fluorescens* and *Bacillus* sp has been a component of integrated disease management and their effect on reducing the impact of disease was appreciable, albeit, the scientific publications on their use are very scanty except academic interest work in form of dissertations and part of the thesis. The inhibitory actions of *Pseudomonas* sp., was reported against anthracnose disease caused by *Colletotrichum* sp., and it is expected to antagonize other foliar pathogens including *Alternaria* sp.. Mycorrhizae associated actinomycete *Streptomyces canus* as a bio-inoculant found promising for improvement of growth and tolerance against bacterial blight. The species of *Trichoderma* like *hamatum*, *harzianum*, *koningii* were found effective *in vitro* against bacterial diseases, however, these are usually soil inhabitants and their efficacy in field conditions at phylloplane has to be evaluated for exploring into commercial purpose.

The BCAs such as *P. fluorescens*, *Bacillus* sp., and various fungal antagonists were found beneficial as most of them are known to release growth promoting agents like gibberellic acid, IAA, cytokinins, auxins, lytic enzymes etc., in plants contributing to improving the biomass, root proliferation, vigour of plants and overall tolerance against biotic and abiotic stresses. Once BCA is introduced and creating the conditions required for the antagonist to establish into thresholds levels, continually keeps the pathogen populations under control with high degree of host specificity (Kabdwall *et al.*, 2018).

Further, use of bacteriophages to manage bacterial blight of pomegranate was also reported as a pathogen targeted approach unlike beneficial microbes, however development of bacteriophage formulations and their application under field conditions is challenging. In addition, copper and their residue are known to have deleterious effect over multiplication of phages. So it demands compatibility assays with other agrochemicals (Li and Dennehy, 2011)

Biorationals for disease management

Active Ingredient	Target Pests
Acibenzolar-S-methyl	Downy mildew, bacterial speck, white rust, *Xanthomonas*
Cottonseed Oil, Corn Oil, Garlic Oil	Powdery mildew
Kaolin	Powdery mildew, sunburn and heat stress
Neem Oil	Anthracnose, *Botrytis*, downy mildew, powdery mildew, scabs, rusts, leaf spots and blights
Oils, Petroleum based	*Alternaria*, Gummy Stem blight, Powdery mildew, Rust
Potassium Bicarbonate	*Alternaria*, Anthracnose, *Botrytis*, downy mildew, Fusarium, Leaf spots, *Phytophthora*, powdery mildew
Potassium Phosphite	Downy mildews, *Phytophthora* species, *Pythium* species
Mono- and dibasic sodium, potassium, and ammonium phosphites	Downy mildews, *Phytophthora* species, *Pythium* species
Monopotassium phosphate	Powdery Mildew
Potassium silicate	Powdery Mildew, Leaf spots
Reynoutria sachaliensis Extract	Powdery Mildew, *Botrytis*, Leaf spots, Bacterial spot, speck and canker
Rosemary and Clove Oils	Powdery Mildews, bacterial spot
Rosemary, Clove, and Thyme Oils	Bacterial Spot, Early Blight, Gray Mold, Late Blight, Powdery mildew, Downy mildew
Rosemary, Thyme, and Clove Oil	Anthracnose, *Botrytis*, downy mildew, powdery mildew, leaf spots, rusts, bacterial spot
Saponins of *Quillaja saponaria*	Nematodes
Sesame Oil	Powdery Mildew, Leaf spots
Sesame Oil	Nematodes
Sesame Seed Meal	Nematodes

Exemplary biorational demonstration for disease management

Chitosan and β-Cyclodextrin-epichlorohydrin Polymer Composite for Carbendazim. This complexing biorational helps to protect against *Sclerotinia sclerotiorum*

Diagrammatical sketch of controlled release of active ingredient targeted for disease management (Wang *et al.*, 2017).

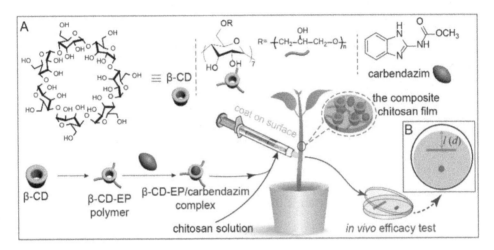

Biopesticides using candidate isolates

Bioagents used for developing the biopesticides	Targeted diseases of horticultural crops
Streptomyces lydicus	Downy mildew, powdery mildew, *Botrytis, Sclerotinia* spp., *Pythium, Phytophthora, Fusarium, Rhizoctonia*
Bacillus pumilus Strain QST 2808	Powdery mildew, rust, *Sclerotinia*, downy mildew, Leaf spots
Pseudomonas syringae Strain ESC-10	*Fusarium* and *Helminthosporium* storage rots
Trichoderma asperellum and *Trichoderma gamsii*	*Fusarium. Rhizoctonia, Pythium, Phytophthora, Sclerotinia, Sclerotium, Verticillium*

Bacillus subtilis GB03	Anthracnose, *Botrytis*, bacterial diseases, Powdery mildew, *Phytophthora*, *Pythium*, *Rhizoctonia*, Leaf spots
Bacillus subtilis QST 713 strain	*Rhizoctonia*, *Pythium*, *Phytophthora*, *Fusarium*
Coniothyrium minitans Strain CON/M/91-08	*Sclerotinia sclerotiorum* *Sclerotinia minor*
Myrothecium verrucaria Strain AARC-0255	Nematodes
Bacillus amyloliquefaciens	*Botrytis*, *Alternaria*, fungal leaf spots & blights, powdery mildew, downy mildew, soilborne pathogens: *Rhizoctonia*, *Pythium*, *Fusarium*, *Sclerotinia*, (suppression only: rust, late blight, early blight),
Streptomyces griseoviridis Strain K61	*Fusarium*, *Alternaria*, *Phomopsis*, *Botrytis*, *Pythium*, *Phytophthora*, *Rhizoctonia*
Bacillus subtilis QST 713 strain	Black spot, Powdery Mildew, Bacterial Spot, Anthracnose, Gray Mold, Downy Mildew, Late Blight
Gliocladium catenulatum Strain Ji1446	*Pythium*, *Phytophthora*, *Rhizoctonia*, *Fusarium*, *Verticillium*, *Botrytis*
Bacillus subtilis QST 713	*Alternaria*, bacterial blight (*Xanthomonas*), downy mildew, powdery mildew, *Sclerotinia* spp., *Botrytis*, rust, *Pytophthora infestans*
Trichoderma harzianum Rifai Strain KRL-AG2	*Pythium*, *Rhizoctonia*, *Fusarium*, *Cylindrocladium*, *Thielaviopsis*
Trichoderma harzianum Rifai Strain T-22 and *T. viriens* Strain G-41	*Pythium*, *Rhizoctonia*, *Fusarium*, *Cylindrocladium*, *Thielaviopsis*
Bacillus subtilis QST 713	*Alternaria*, Anthracnose, *Botrytis*, downy mildew, powdery mildew, rust, *Helminthosporium* diseases, *Didymella*, *Phoma*, Bacterial diseases

Bacillus pumilis QST 2808	*Alternaria*, Downy mildew, Powdery mildew, Rust
Bacillus subtilis MBI 600	*Rhizoctonia, Fusarium, Pythium,* Powdery mildew
Bacillus subtilis var. amyloliquefaciens FZB24	*Rhizoctonia, Fusarium, Sclerotinia, Pythium, Phytophthora,* leaf spots, Powdery mildew
Bacillus amyloliquefaciens D747	*Botrytis, Rhizoctonia, Fusarium, Sclerotinia, Pythium, Phytophthora,* leaf spots, powdery mildew, downy mildew

Biological formulations for management of disease in horticulture crops (Manjunath G and Sharma, 2015)

Formulations	Organism associated	Manufacturer	Remarks/claims
Probiotic Growbax	An effective combinations of naturally occurring beneficial microbes	Microbox India Limited, Hyderabad	Enhances resistance against pests and pathogens.
Probiotic Sulfex	*Bacillus*	Microbox India Limited, Hyderabad	Enhances immunity
Sudo	*Psedomonas fluorescence*	Kan Biosys Pvt Ltd	Effective against a wide variety of seed and soil borne plant pathogenic fungi
Nemagon	*Paecilomyces lilacinus*	Green life Biotech Laboratory	Attacks nematodes and their eggs without being harmful to any plant species

Xoton Plant immunizer	-	West cost Rasayan International Pvt. Ltd	Increases immunity of plant to fight against phycomycetes fungti like downy mildew, late blight, quick wilt and root rot. rtc.
Top life Super smart-c	See weed extract	ROOTS- Florentine Curtis. Pte. Ltd. Singapore and Nashik Marketed by; TL Agro. Pvt. Ltd. Toplife House, Pune, Maharastra	Increases immunity
Phytoalexin 84	Herbal extracts	Akshay Chemicals, Ratnagiri	Increases immunity
Kil Tel	Chitosan based	Camlin Fine Sciences Ltd. Mumbai	Improves plant growth and defense mechanism
Zanthonil	*Oscimum sanctum*, ginger oil, clove oil	Sanardhini Agro. Pvt. Ltd. Satara	Defense booster
Kalisena SA	*Aspergillus niger*	Cadila Pharmaceuticals, Ahmadebad	Induction of resistance, control of soil borne wilt pathogens, improved growth and production
Josh Super	Vesicular Arbuscular Mycorrhizae	Cadila Pharmaceuticals, Ahmadebad	Better rooting and enhanced growth

Summary

The use of biopesticides/biorationals under the ambit of IDM enables the effective management of diseases with reducing load of toxic pesticides and their residues in horticulture crops. However, if we compare the status of biopesticides/biorationals use *vis-a-vis* with other molecules situations, the picture is not satisfactory. A regulation of biopesticide quality is a matter of high concern to ensure the better BCAs with longer shelf life and appropriate delivery methods are approved for their continuous use in IDM. The re-isolation of applied antagonists from plants foliage has little chances of recovery when it is used through talc formulations. Therefore, emulsifier based formulations are needed for sustaining activity of BCAs temporally with long persistence. Further, the efficacy of biopesticide products have to be assessed periodically once after the release into farmers fields. These lacunae can be overcome by using biorationals largely. Further, greater cooperation with industrial scientists is required for transferring the identified technology into mass-scaling for serving the products on a large scale.

References

- Kabdwal B. C. Roopali Sharma Rashmi Tewari Anand Kumar Tewari Rajesh Pratap Singh Jatinder Kumar Dandona, 2018. Field efficacy of different combinations of *Trichoderma harzianum, Pseudomonas fluorescens* and arbuscular mycorrhiza fungus against the major diseases of tomato in Uttarakhand (India). *Egyptian Journal of Biological Pest Control.29:1-15*

- Li J1 and Dennehy J.J., 2011. Differential bacteriophage mortality on exposure to copper. *Appllied Environment Microbiology. 77(19):6878-83.*

- Manjunatha, G. and Sharma, J., 2015. Trends and perspectives of using biopesticides in pomegranate for improved tolerance against wilt and blight. Proceedings published by National conference of Pomegranate, Solapur. Dec.5-7,2014

- Muhammad Sarwar, 2015. Usage of Biorational Pesticides with Novel Modes of Action, Mechanism and Application in Crop Protection. *International Journal of Materials Chemistry and Physics. 1:156-162*

- Wang, Peibin Zhang Fan, HuYifan Zhao and Liping Zhu. 2017. A crosslinked β-cyclodextrin polymer used for rapid removal of a broad-spectrum of organic micropollutants from water. *Carbohydrate polymers .177:224-231*

3

Trichogenic Membrane Nanoemulsion Formulations - Novel Biopesticides for Crop Disease Management

N. Geetha

University of Mysore, Mysuru

Nanotechnology is a multi-disciplinary approach which involves formation and utilization of different methods on a nanometric scale. Numerous types of nanoformulations have been reported, including nanoemulsions, which are dispersed systems composed of immiscible liquids with one or more stabilizers. Nanoemulsions of lipids are oil-in-water dispersions with droplet diameter in the range of 10-100 nm (Mc Clements, 2011). Nanoemulsion is one of the most accepted formulation because of its improved bioavailability, optical transparency and greater physical stability (Huang *et al.*, 2010; Karadag *et al.*, 2013; Peshkovsky *et al.*, 2013). Physico-chemical properties of nanoemulsions are influenced by several features, such as oil type, surfactant and process conditions (Einhorn-Stoll *et al.*, 2002; Mc Clements, 2011). Additionally, low viscosity, high kinetic stability and optical transparency formulate them as a very attractive system for various industrial applications for example, in agrochemicals for pesticide delivery, in the pharmaceutical field as drug delivery systems (Wu *et al.*, 2001), in the chemical industry as polymerization reaction media and in cosmetics as personal care formulations (Sonneville-Aubrun *et al.*, 2004).

The application of nanotechnology in agriculture aims in particular to reduce applications of plant protection products, minimize nutrient losses in fertilization and increase yields through optimized nutrient management. Nanotechnology devices and tools, like nanocapsules, nanoparticles and nanoemulsions are used for the detection and treatment of diseases, the enhancement of nutrients absorption by plants and the delivery of active ingredients to specific sites.

Nanomaterials with unique chemical, physical, and mechanical properties e.g. electrochemically active carbon nanotubes, nanofibers etc have been recently developed and applied for highly sensitive bio-chemical sensors. These nanosensors have also relevant implications for application in agriculture, in particular for soil analysis, easy bio-chemical sensing and control, pesticide and nutrient delivery.

Numerous types of nanoformulations have been reported, including nanoemulsions, which are dispersed systems composed by immiscible liquids with one or more stabilizers. Nanoemulsions of lipids are oil-in-water (O/W) dispersions with droplet diameter in the range of 10-100 nm. Nanoemulsion is one of the most accepted formulations because of its improved bioavailability, optical transparency and greater physical stability. Physico-chemical properties of nanoemulsions are influenced by several features, such as oil type, surfactant and process conditions.

Nanoemulsions involve input of mechanical energy and low energy emulsification process and are also prepared by high-energy emulsification. The high-energy emulsification process includes high-shear blending, high-pressure homogenization and ultrasonication. Ultrasonic emulsification is one of the high energy approaches to develop nanoemulsion. This method is a proficient method for developing stable nanoemulsion with small droplet size and low polydispersity index. It uses high frequency sound waves to cause mechanical vibrations and formation of cavity. Pesticide formulations with O/W nanoemulsion and its efficacy was equivalent to microemulsion formulation. Studies on *Trichoderma* and its membrane lipid nanoemulsion formulation help us to understand the host-pathogen interactions and mechanisms involved in pathogen associated molecular patterns (PAMPs). Seed priming is a process, which occurs through absorption of the elicitor molecule. So, the compounds used for seed priming process must be solubilized and dispersed in water. Oils are volatile complex mixtures with a wide range of biological activities, including repellent, insecticidal and larvicidal properties. With this context, an O/W nanoemulsion of extracted crude lipid from *Trichoderma* spp. overcomes solubility in water. Factors such as surfactant and emulsification properties significantly influence the formation of emulsion. Further, synthesized lipid emulsion of *Trichoderma* spp.

was analyzed for its physico-chemical characteristic properties. Lipid emulsion obtained after sonication for 1h was subjected to particle size analysis by zeta potential and dynamic light scattering process (DLS). It was found that particle size in emulsified sample was found in the range, 5-51 nm, with polydispersity index (PDI), 0.156 to 0.333. The formation of emulsion and its surface methodology signifies the influence of the lipid, emulsifier concentration, emulsification time and formation of droplet size. Generally, smaller the droplet size signifies greater the stability of the emulsification.

The objective of the present study was to formulate nanoemulsion of membrane lipid extracted from different *Trichoderma* spp. and to evaluate efficacy of the synthesized nanoemulsion for the induction of disease resistance in pearl millet against downy mildew pathogen. The formulated nanoemulsion of membrane lipid extracted from different *Trichoderma* spp. was evaluated for the efficacy to induce disease resistance in pearl millet against downy mildew pathogen. The interaction between the pathogen and host induces challenges in cell metabolism primarily in the enzyme activities including the jasmonic acid signaling intermediates. In the membrane lipid nanoemulsion treatment, significant ($P \leq 0.05$) disease protection was achieved from *T. brevicompactum* against downy mildew pathogen. Further, time interval studies using the best treatment were carried out to study the induction of resistance.

In this context, our results suggest that the nanoemulsion containing 5% (w/w) of crude membrane lipid of *Trichoderma* spp., 5% (w/w) of polysorbate 80 (w/w) and 90% (w/w) of water can be considered a promising biocontrol formulation for the control of the Oomycetes pathogen, *Sclerospora graminicola* of pearl millet. Glycosphingolipids isolated from various species of *Fusarium oxysporum* also induced resistance against some *Fusarium* diseases. Extractions and purification of cerebrosides from different fungi has been reported earlier.

An ultrasonic method was preferred compared to others because of its low cost equipment, smaller footprint, easy cleaning with good service (Peshkovsky *et al.*, 2013). Excessive surfactant may result in lower diffusion rate of surfactant and the lower content might lead to a lower diffusion and amalgamation of the emulsion droplets (Li and Chiang, 2012). Lipid content and ultrasonication time influences the droplet size at a fixed surfactant content of 5%. Stability of emulsion was significantly ($P \leq 0.05$) influenced by the ultrasonication time interval, as it effects droplet size distribution and surfactant adsorption rate on the droplet surface (Li and Chiang, 2012).

In emulsion, sedimentation is a reversible destabilization phenomenon, whereas variation of the droplet size is an irreversible process (Abismail *et al.*, 1999). Polydispersity index (PDI) was found in the range of 0.156 to 0.333 that means distribution of the droplet size is in the narrow range. Further, zeta potential was found in the range of +17.8 mv to -28.2 mv and pH of 7. Optimal nanoemulsion with low turbidly (600 nm absorbance = 0.36 ± 0.01) signifies the smaller droplet size (Leong *et al.*, 2009). This fact makes this nanoemulsion suitable for its incorporation into different systems without altering visual quality. Furthermore, there was no significant difference noticed in polydispersity during the incubation period. Micelles are continuously disintegrating and reassembling, being in dynamic equilibrium with individual surfactant molecules. Optimum condition provided facilitate the formation of the stable O/W emulsion with minimum droplet size i.e., within the range of nanometer. In wheat bran emulsion, it was observed that surfactant to oil ratio was relatively high. Li *et al.* (2012) demonstrate the nanoemulsion-based delivery systems for poorly water-soluble bioactive compounds. Recently, Naveen *et al.* (2013) studied the cerebroside-mediated disease resistance in Chilli through eliciting the production of defense-related enzymes, accretion of H_2O_2 and accumulation of capsidiol.

The results of the present study suggest that droplet size of membrane lipid nanoemulsion of *Trichoderma* spp. is inversely proportional to the induction of disease resistance in pearl millet. Further, membrane lipid nanoemulsion from *T. brevicompactum* is much effective in reducing downy mildew disease of pearl millet. Therefore, this study points to the potential integrated approach in the future management of downy mildew disease in pearl millet. The result of the present study illustrates potential efficacy data of lipid nanoemulsion of *Trichoderma* spp. to control of downy mildew pathogen. Additionally, it opens a new avenue, where *Trichoderma* membrane lipid nanoemulsion formulations can be successfully employed for plant disease management.

References

- Einhorn-Stoll, U., Weiss, M. and Kunzek, H., 2002. Influence of the emulsion components and preparation method on the laboratory scale preparation of o/w emulsions containing different types of dispersed phases and/or emulsifiers. *Molecular Nutrition and Food Research,* **46** (4): 294-301.

- Karadag, A., Yang, X., Ozcelik, B. and Huang, Q., 2013. Optimization of preparation conditions for Quercetin Nanoemulsions using response surface methodology. *J. Agric. Food Chem.,* **61** (9): 2130-2139.

- Li, T. and Chiang John, Y. L., 2012.Bile Acid Signaling in Liver Metabolism and Diseases.*Interactions between Bile Acids and Nuclear Receptors and Their Effects on Lipid Metabolism and Liver Diseases.* https://doi.org/10.1155/2012/754067

- McClements,2011, Edible nanoemulsions: fabrication, properties, and functional performance soft matter.https://doi.org/10.1039/C0SM00549E

- Peshkovsky, A., Peshkovsky, S.L., and Bystryak, S., 2013. Scalable high power ultrasonic technology for the production of translucent nanoemulsions. *Chemical Engineering and Processing: Process Intensification,* **69:** 77-82.

- Sonneville-Aubrun, J., Simonne, T. and Alloret, F. L., 2004.Nanoemulsions: a new vehicle for skincare products. *Advances in colloid and interface Science,* 108: 145-149.

- Wu, S. Z., Harish, S., Sanders-Millare, D. and Guruwaiya, J. A., 2001. Implantable medical device having protruding surface structures for drug delivery and cover attachment. United States Patent US6254632B1

4

Bottlenecks for the Promotion and Adaptation of Biopesticides in India

Yogesh D.[1] and Raghavendra M.P.[2*]
[1,2]University of Mysore, Manasagangotri, Mysuru
[2]Maharani's Science College for Women, Mysuru

Abstract

Living organisms are susceptible to diseases caused by bacteria, protozoa, fungi, nematodes, viruses and other sub viral particles. This is the foundation for exploiting the living organisms which can kill both phyto and human pathogens bothering human beings. They are referred to as biopesticides and these are gaining momentum due to its selective action, natural origin and minimal or no effects on the non target organisms. World is yet to witness the second green revolution and in this connection biopesticides remain as an important component of the integrated pest management for sustainable agriculture. Even though biopesticides seems to be better alternative to the chemical pesticides which is widely accepted and easily practiced in conventional agricultural practices, it has to pass through several hurdles to replace or reduce the use of chemical pesticides having several issues concerning environment and health. This chapter intends to discuss the issues related to its benefits and problems associated with production and field performance.
Keywords: Benefits; future challenges; goals; sustainable agriculture

Introduction

With the development of agriculture, humans are practicing various methods to control pests. Hence, agriculture and pest control are contemporaries to mankind. There is fight for food between each and every organism, even humans and pests compete for food either at agricultural fields or during its storage. One needs to protect their share of food through one or the other means and hence is the concept of food protection.

For more than a century, use of chemicals for control of pests is a very common practice in agriculture. With the advancement in chemistry of pesticides various synthetic pesticides have been introduced. However, these were with the limitations of toxicity to human as well as environment. Apart from these the major concerns over its applications is its effect on soil, environment and the residue in food products along with the problems associated with the increase of resistance in the pests.

In view of these, it is very much necessary to replace chemical pesticides with better and safer alternatives. In this regard bio-pesticides are best alternative due to its biological origin with least or no toxicity to humans and animals (Samada and Tambunan 2020).

Application of biopesticides in pest control is now the preferred alternative over the use of synthetic pesticides. The pest control ability as well as diverse mode of actions of biopesticides helps in evading the resistance development in the pests. India being agrarian country, a huge diversity of flora and fauna is observed and hence there is an urgent requirement for identifying new and effective biopesticides which help in managing pests associated with diverse cultivated crops.

At present, strains of *Bacillus thuringiensis*, fungi, nuclear polyhedrosis virus (NPV) and parasitic nematodes have been commercially exploited as biocontrol formulations. These biopesticides as biocontrol agents emerges to be more promising advances to address the issues related to phytopathogens as well as environmental problems that are resulted due to extensive use of synthetic pesticides.

Biopesticides in Present Scenario

Among the pesticides consumed in India, insecticides accounts for more than 50%, followed by herbicides and fungicides. With the limited number of chemical pesticides in the Indian market and due to the development of resistance in pests as well as observed health hazards linked with use of synthetic pesticides have increased the demand for biopesticides. Based on the reports of Department of Agriculture Cooperation and

Ministry of Agriculture and Farmers Welfare, it was estimated that the consumption of biopesticides in India has increased from about 219 tonnes during 1996–1997 to 683 tonnes in 2000– 2001, it was around 3000 tonnes in 2015–2016 and the biopesticide market in India was USD 70.45 million in 2016 and it is growing with a CAGR of 17.08% (https://www.mordorintelligence.com/industry-reports/indian- biopesticides-market).

Limitations and Hurdles for Biopesticides

Following are some of the key limitations and hurdles for biopesticides.

Un-availability of biopesticides

The greater dependence of Indian farmers on use of synthetic chemical pesticides can be reduced or removed only by substantial increment of biopesticide industry. There are various factors that have been hindering the growth potential of biopesticides market in India. The major factors are unavailability, lesser reach and also vanishing of multiple or mixed cropping system.

Inadequacy of quality of biopesticides

There are several studies conducted among which the study by NBAIR revealed that about 50–70% of microbial biopesticide-based products available in India are with the issues such as fewer colony propagules in the formulations than that is listed on the label, solid formulations with excessive moisture content in them, or presence of contaminants. These formulations failed to meet CIB and RC standards (Ramanujam *et al.*, 2014). Survey carried out by the National Pesticide Manufacturers Association (NAPM) revealed that many biopesticide formulations sold in the market have no clear identifiable registration number of the company or details of addresses and also active ingredients that need to be mentioned on the label (FICCI, 2015).

Limited shelf life of biopesticides

Limited shelf life comes next to the inadequate quality as another constraint for many microbial biopesticides. This is prevalent especially in most parts of rural India where there is limited access to fresh biopesticides formulations and also inadequate facility for storage such as refrigeration aggravated by limited power supplies (Mishra *et al.*, 2015). Many dry formulations of entomopathogenic fungi and nematodes have been claimed to have a shelf life of up to a year when kept under refrigerated conditions and due the lack of proper storage facilities many liquid formulations of biopesticides become ineffective (Ramanujam *et al.*, 2014). When it comes to effectiveness, their slow rate of killing, limitations of field persistence due to high levels of UV radiation becomes a major shelf life issue. For example, exposure to UV light known to inactivate many entomopathogens and there is a need for development of formulation technology which can protect the inoculum and enhance the exposure to the target pest organisms in the field conditions. Poor water solubility of some preparations poses additional challenges in development of commercially workable microbial pesticides (Aneja *et al.*, 2016).

Slow process and un-predictable stability and inconsistent field results

Another major hurdles for using biopesticides is that cost, when compared with chemical pesticides, microbial pesticides are expensive, slower in action and the results of field efficacy is inconsistent. In general farmers prefer inexpensive as well as faster pest management options. Biopesticides are slower in action for instance; it takes about 3.5 days for target organism to die from entomopathogenic biopesticides. Another key limitation is a wider variation with respect to mortality achieved based on the crop, pest and also the environment.

Constraints for commercialisation of microbial biopesticides in India

Even though microbial biopesticides or insecticides offers several advantages compared to traditional products for pest control, they are yet to achieve a wider range of commercial application in India. There are several factors which limits the microbial pesticide in Indian market. As mentioned earlier, the basic limitation starts with quality of the product, quality control includes low counts of microorganisms that need to be used as biopesticides which results in under performance in the field. The other limitations are lack of production at larger scale and the trade products which are not registered in market (Alam, 2000; Arora *et al.*, 2010; Gupta and Dikshit, 2010; Mishra *et al.*, 2015).

Other major constraints for the promotion of biopesticides

Lack of profit– Large number of biopesticides are highly selective in nature. The baculoviruses based bioinsecticides such as the CpGV are typically selective in action against a particular insect; hence, they have low potential with respect to profit as the size of their market is small or limited.

Fixed costs – Farmers who use biopesticides face a large fixed cost compared to those who use chemical pesticides. This is because the fixed cost associated with biopesticides which is not distributed among farmers at large, and it is disadvantageous for the early users.

Suspicions or uncertainties for the use– There are lot of suspicions or uncertainties in the minds of farmers when it comes to use of biopesticides as a new product about which they lack practical experience. This is due to the fact that, for a very long time large number of farmers have been using conventional chemical pesticides and have gained sufficient substantial experience as well as confidence in their effectiveness.

Variability with respect to performance

Farmers prefer consistency, reliability and efficacy while selecting a pest control method. The economic feasibility will increase as biopesticides become more consistent with respect to their performance in field conditions.

Cost of biopesticides

The costs associated with respect to labour, time and management, when using a new product or technique can raise the overall cost (Dar *et al*.,2010). In addition, the higher price of biopesticides have limited their usage by poor Indian farmers which devalue the superiority and advantages of these biopesticides in Indian market.

Malpractices in biopesticide market

Certain malpractices are the cause which have rendered biopesticides ineffective. This is in case of some markets around the world with no stringent regulation systems is in place as in case of developed countries. Adulteration using inert and ineffective material, selling of expired formulations in new packaging, improper storing conditions, or even incorporating chemical pesticides with biopesticides damages the status of biopesticides. These limitations, however, outshining the potential of biopesticides and their successful usage in pest management (Dar *et al*.,2010).

Hurdles of Regulatory/Statutory Processes

Ultimately, the strict, expensive as well as time-consuming registration process in our country slows down the microbial biopesticides development. At present, the time duration between granting of patents and registration of biopesticide formulations exceeds time duration beyond 5 years (Venkatesan and Pattar, 2017). This appears to be one of the reasons which aggravate the trading of several unregistered products of poor quality.

Importance of Indigenous Microorganisms as Biopesticide

Environmental protection is the primary importance in the present day life of mankind. Researchers have been searching for ecofriendly technologies for betterment of agriculture, waste management and disease control etc. Indigenous Microorganisms (IMO's)-based technology is one such important technology which is applied in many parts of the world for processes such as minerals extraction, agriculture enhancement, waste management and pest control. Indigenous microorganisms are basically a consortium of innate microorganisms which inhabits the soil as well as the surfaces of living organisms both inside and outside. These have the capabilities in biodegradation, biocomposting, nitrogen fixation, biocontrol, production of plant growth hormones and also improve soil fertility. The environmental restoration and conservation with the use of indigenous microbes in a native manner into productive as well as protective bioresources is the primary concern in this chapter. Based on the site of sample collection, the process and methods of isolation biopesticides are need to be monitered as they may differ from region to region. Eventually, one can address the gap existing between the environmental distress and the hostile activities which is continuously provoked by anthropogenic activities by employing indigenous microorganisms into action.

IMOs being indigenous have the potential to make environment favorable to improve and maintain soil flora, fauna and the other microorganisms, which in turn support the higher plants, animals including humans. The IMOs are environmentally safe, eco- friendly as well as healthy with great potential to create hunger-free environment and better quality of crops and livestock which is assured even in absence of chemical fertilizers and pesticides as inputs (Kumar and Gopal, 2015).

In recent studies, the use of indigenous versus synthetic microbiomes in controlling soil-borne diseases was investigated (Mazzola and Freilich, 2017). It was observed that there was clear advantages with respect to the ability to survive and efficacy when microorganisms adapted to the specific environmental condition or competition for a similar niche in the phytobiome are used. Numerous examples are available in literature about the benefits of using indigenous microflora as biocontrols. For example, fungi antagonistic to *Rhizoctonia* was identified and has been shown to bring down the severity of disease on numerous crops. Of these, some are nonpathogenic *Rhizoctonia solani,* binucleate *Rhizoctonia* and fungal members of other genera include *Trichoderma* spp. It was also reported that *R. solani* AG11 is associated with reduced soybean seedling disease which is caused by a *Pythium* spp. (Spurlock *et al.,* 2016).

On the other hand, Failure of most of the microbial biopesticides may be due to their adaptability. Nativity of the biopesticides, the habitat from where it is isolated and efficient experimental procedures which can prove the efficacy of the biopesticide in different environment or soil types are need of the hour. Technology need to be developed to isolate biopesticidal microorganisms from soil which is indigenous to particular type of soil or environment, rapid identification of microorganisms and molecules or genes responsible for their efficient biopesticidal activity, rapid scale up and adaptability for the environment and ease of application.

Role of Indigenous Microorganisms for Sustainable Environment

These microbes suppress pathogen populations associated with the disease through number of mechanisms. Many studies have documented the capabilities of microbial biopesticides to contain both root and foliar diseases (Emmert and Handelsman, 1999; Doumbou *et al.,* 2001; Fravel, 2005; El Tarabily and Sivasithamparam, 2006). However, the general agreement among several studies is that integration is the key to attain consistent activity from biopesticides (Jacobsen *et al.,* 2004; Fravel, 2005). Amongst the microorganisms that are used for biological control of many plant diseases, members of the genus *Bacillus* in particular have been exploited widely. This is due to the characters such as their long duration stability as well as its antimicrobial activity (Fravel, 2005). Suppression ability of these has been attributed to antagonism of pathogen growth through the production of various byproducts of metabolism (Peypoux *et al.,* 1999; Bonmatin *et al.,* 2003). In addition, it is also demonstrated that some of the metabolites from these are also known to stimulate plant defenses and presents an additional cover of control (Ongena *et al.,* 2007). Many species of *Streptomyces* have also been used for disease control against fungal pathogens.

Molecular methods for accurate detection of biopesticides

Natural host-pathogen interactions are manipulated to the profit of humans, crop protection in agriculture as well as forests and to manage diseases from phytopathogens. The isolation and identification of an insect pathogen and its phylogenetic analysis are the primary requirements in plant pathology. Complete sequencing of genomic DNA and analysis are used to assess diversity and phylogenetic analysis of many entomopathogens which can be used as biopesticides (Gani *et al.*, 2019). As an alternative approach, specific genes can be targeted for sequencing.

Sequence analysis of single gene has been largely studied to assess phylogenetic relationships of known as well as novel biocontrol microorganisms. Yet, the lack of adequate resolution and discrepancies with other gene, phylogenies has led to search for other genes and methods to further investigation on evolutionary relationships of entomopathogenic microorganisms.

Fungi and bacteria are the key members of biopesticide formulations used to control pests and among biopesticides, the use of fungi is increasing because of its species diversity. The popularity is due to their well-known hosts, many fungi species can be cultured in artificial nutrient media and their suitability for commercial production. Especially in biological control formulations members of genus *Trichoderma*, such as *Trichoderma koningii*, *Trichoderma harzianum*, *Trichoderma virens*, *Trichoderma viridae* and *Trichoderma pseudokoningii* are used successfully (Tozlu *et al.*, 2018).

Molecular studies may help in revealing any variation which is present among the isolates in support of the diversity in their physiological characteristics. It is well- known that prior knowledge about the relationships at genetic level among breeding materials is essential for the efficient usage of the germplasm for breeding program. Researchers have attempted for genetic analysis of different *Trichoderma* isolates from different geographical locations. Even though no specific markers were found to be effective in discriminating different isolates effectively, use of ISSR technique helped in revealing some degree of polymorphisms with respect to variation among different *Trichoderma* isolates. The basic idea of the study was to determine genetic variations among the isolates of *Trichoderma* spp. primarily for evaluating their efficiency as biopesticides (Shahid *et al.*, 2014).

The polymorphisms for variation study of thirty five *Trichoderma* isolates was performed using ISSR marker and the results of the experiment indicated significant molecular variation among the isolates with respect to morphological characters. The study infers that ISSR or microsatellite analysis can be effectively used for differentiating *Trichoderma* species. There is a need for molecular method which can be used to distinguish specific and effective *Trichoderma* species employed as biocontrol agent from other

Trichoderma species. This can help in quality control, authentication of biocontrol formulations and in avoiding less efficient products sold as effective biocontrol in the market. It holds good for other bioformulations used as biopesticides in agricultural fields.

The extensive and wide studies of genome on gene content, gene expression, conserved gene order, regulatory networks, functional genome annotation, metabolic pathways can all be improved by evolutionary studies which are based on phylogenetics. Hence, the molecular criteria and tools of bioinformatic for pathogen discrimination as well as species demarcation are obvious and need of the hour.

Molecular phylogeny is sensitive to the assumptions and the particular models employed for trees constructing. These things face issues associated with long-branch attraction, saturation and taxon sampling. This implies that strikingly variable results can be obtained by applying different models for the same set of data. Development of new experimental approaches to identify the links between the molecular mechanisms and ecological processes can considerably improve inference in evolutionary biology in a significant way (Gani *et al.*, 2019).

The Random Amplified Polymorphic DNA (ISSR) technique (Williams *et al.*, 1990; Welsh and McClelland, 1990) which involves simultaneous amplification of several unknown loci in the total genome employing primers of arbitrary sequence has been utilized for genetic, taxonomic as well as ecological studies of several fungal members including *Trichoderma* (Paavanem-Huhtala *et al.*, 2000). The ability to reliably differentiate members of different species, fingerprint of various genotypes and calculation of the magnitude of variation within a species is helpful for a breeding program. ISSR is an easy and inexpensive key molecular method used for such purposes. The advantages of this ISSR technique is that it requires a small amount of DNA of about 5-20ng, single short primers of arbitrary sequence (9 to 10 bp), and the rapidity in screening for polymorphisms, the efficiency in generating a large number of markers for genomic mapping and the automation of the technique. There is no requirement of prior knowledge of sequence (Sobral and Honeycutt, 1993). Here in this technique the primers is arbitrary and hence any organism can be mapped using the same set of arbitrary primers. However, there is some loss of information as ISSR markers are dominant rather than co-dominant. If one of the alleles present at an ISSR site is not amplifiable, then marker homozygote cannot be distinguished easily from marker/null heterozygote. Other problems may arise if the products of different loci have similarity in molecular weights and so it will be difficult to distinguish on a gel due to comigration. The uncertain homology problem appears to be serious at higher taxonomic levels where only a few shared bands are produced. Even though ISSR technique has some limitations, it is one of the most powerful techniques in genetic studies and variation in microorganisms (Mailer *et al.*, 1994) and for the construction of the first linkage maps for many plant and pathogens (Yang and Quiros, 1995).

Shahid *et al.*, (2014) have used six ISSR primers for testing the polymorphism and found genetic diversity greater than 90% among the *Trichoderma* spp. This indicates that there is complete variability within the species isolated from different fields. The studies have shown that *Trichoderma* species have very good diversity and there is strong possibility to develop the isolates specific primers which can be employed for identifying the particular *Trichoderma* isolate with better biological potential directly from the field isolates bypassing the cumbersome bio assay.

Researchers have studied Inter- and intraspecific variation in the entomophthoralean genera *Pandora*, *Entomophthora*, *Zoophthora*, and *Entomophaga* using amplification of the ITS region, RFLP and RAPD, which have indicated the existance of polymorphisms within as well as between the species (Hodge *et al.*, 1995; Hajek *et al.*, 1996; Rohel *et al.*, 1997, Sierotzki *et al.*, 2000; Jensen and Eilenberg, 2001; Nielsen *et al.*, 2001). Earlier studies have compared the sizes of ITS regions and have provided some valuable information on groupings within *P. neoaphidis* (Rohel *et al.*, 1997, Sierotzki *et al.*, 2000, Nielsen *et al.*, 2001).

SCAR (Sequence Characterized Amplified Regions) Marker

Parallel with the isolation of new and highly-active strains, there is scope to develop studies on the requirements of selected strains for nutrient substances, precursors, investigations of optimal aeration conditions, temperature conditions for field performance, methods of preparation and growth on plating material and so on. Increase in activity of microorganisms, genetics plays the most important role. It is quite obvious that success in the selection of microorganisms depends on the application of genetic methods. Numerous attempts to ignore genetic methods in the selection of microorganisms have always led to failure.

Restriction fragment length polymorphism (RFLP) analysis has been found useful to determine the taxonomic relationships among fungi. This technique makes use of restriction enzymes to digest DNA, which is then separated by agarose-gel electrophoresis. Difference in the size and number of restriction fragments can be detected by southern blot analysis or be observed directly by staining gels with ethidium bromide. These differences can result from loss or gain of restriction endonuclease recognition sites. Depending on the restriction enzyme used, digestion of target DNA may produce a set of specific fragments that can be considered as a fingerprint for a given strain.

Due to the sensitivity of RAPD technique to PCR conditions, there is less reproducibility of RAPD results. This problem was solved by Paran and Michelmore (1993) who first documented the use of SCAR (Sequence Characterized Amplified Regions) marker in lettuce, where the marker was related to downey mildew resistance genes. SCAR markers can be developed by producing primers from unique polymorphic RAPD band. The primer is then allowed to amplify the genomic DNA of any particular genotype and not of other genotypes. Thus, PCR conditions of RAPD technique can be made more reliable by using SCAR markers.

Future Prospects of Biopesticides

In spite of many challenges, scientific community believes that the biopesticide market in India has a bright future. Biopesticide research is at a relatively early developmental stage in India but is evolving rapidly. With the increasing research focus on identification of effective indigenous microorganisms, improvising formulation, mass production, adoption of technologies can reduce the costs and enhance the shelf life of the products and there is a need for exploration of additional markets.

Even it demands reforms in current registration process for biopesticides and update on toxicological data of the products. The dossier submission requirements for microbial biopesticides should be streamlined which can reduce costs and accelerate process of registration. Similarly, the patenting process for microbial products needs to be addressed by legislation. There is also a need for review regarding the effectiveness of current CIB&RC policies accessibility as well as affordability of biopesticides, the quality-control issues of the products and the sales of unregistered products (FAO and WHO, 2017; Kumar et al., 2019).

Conclusion

Growth of the biopesticide market has been growing at double digits globally, which may be due to strong demand for organic food in developed markets. USA is a major consumer of the global biopesticides production while in India, the uptake has been relatively slow and biopesticides have low market share. Many fungi and bacteria such as *Trichoderma* strains, *Pseudomonas fluorescens, Bacillus thuringiensis* along with neem derived products, dominate the biopesticide market.

For risk adverse farmers, who have internalized practice of pest control and management with synthetic pesticides, shifting to biopesticides would be a major challenge as they have to re-learn the entire process in which their income is at stake. Earning the trust and proving efficacy and performance has been a greater challenge for biopesticides.

Around 500 biopesticides are available in the Indian markets which have been duly registered by the CIB&RC. These products have been demonstrated for their efficacies in many laboratories; however, problem of their quality control needs to be addressed.

One of the major hurdles for the promotion biopesticides as effective alternative to chemical pesticides is the lack of profile for biopesticide, which indicates the weakness of the encouraging policy network. Relative crudeness of the policy network, resources limitation and capabilities, and lack of trust between producers and regulators are some of the serious issues associated with this. Better knowledge and understanding of the biopesticides and their mode of action, effects and regulatory issues which arises in their adoption may aid further to increase their profile among the public, policy-makers and helps them to realize their role on sustainability. The environmental safety is a global concern, one needs to increase awareness among the farmers, manufacturers, government bodies, policy makers and the common people to switch-over to biopesticides for the pest management needs.

This chapter calls for a shift in paradigm with respect to total dependence on synthetic pesticides to biopesticides and stress on research, development, promotion and use of indigenous biopesticides as a sustainable, eco-friendly and effective alternative for the alleviation of pests and diseases.

References

• Aneja, K.R., Khan, S.A. and Aneja, A., 2016. Biopesticides an eco-friendly pest management approach in agriculture: status and prospects. *Kavaka,* **47**:145–154.

• Bonmatin, J.M., Laprevote, O. and Peypoux, F., 2003. Diversity among microbial cyclic lipopeptides: iturins and surfactins. Activity-structure relationships to design new bioactive agents. *Combinatorial Chemistry and High Throughput Screening,* **6**: 541–556.

• Dar, S.A., Khan, Z. H., Khan, A.A. and Ahmad, S.B., 2010. Biopesticides– Its Prospects and Limitations: An Overview. *Perspectives in Animal Ecology and Reproduction,* 22: 296-314

• Doumbou, C.L., Hamby Salove, M.K., Crawford, D.L. and Beaulieu, C. 2001. Actinomycetes, promising tools to control plant diseases and to promote plant growth. *Phytoprotection,* **82**: 85–102.

• El-Tarabily, K.A. and Sivasithamparam, K., 2006. Non- streptomycete actinomycetes as biocontrol agents of soil-borne fungal plant pathogens and as plant growth promoters. *Soil Biology and Biochemistry,* **38**: 1505–1520.

• Emmert, E.A.B. and Handelsman, J., 1999. Biocontrol of plant disease: a (Gram) positive perspective. *FEMS Microbiology Letters,* **171**: 1–9.

• FAO and WHO, 2017. International Code of Conduct on Pesticide Management: Guidelines for the Registration of Microbial, Botanical and Semiochemical Pest Control Agents for Plant Protection and Public Health Uses. Food and Agriculture Organization of the United Nations World Health Organization, Rome, p. 2017.

• FICCI (Federation of Indian Chambers of Commerce and Industry), 2015. Study on substandard, spurious/counterfeit pesticides in India, p. 66. https://croplife. org/ wpcontent/ploads/2015/10/Study-on-sub-standard-spurious-counterfeit- pesticides- in- India.pdf.

• Fravel, D.R., 2005. Commercialization and Implementation of Biocintrol1. *Annual Review of Phytopathology,* **43**: 337–359.

• Gani, M., Hassan, T., Saini, Gupta, R.K. and Bali K. 2019. Molecular Phylogeny of Entomopathogens. In: M. A. Khan, W. Ahmad (eds.), Microbes for Sustainable Insect Pest Management, Sustainability in Plant and Crop Protection, Springer Nature Switzerland AG p. 43-113.

- Hajek, A. E., Hodge, K. T., Liebherr, J. K., Day, W. H. and Vandenberg, J. D., 1996. Use of RAPD analysis to trace the origin of the weevil pathogen *Zoophthora phytonomi* in North America. *Mycological Research,* **100**: 349–355.

- Jacobsen, B.J., Zidack, N.K., Larson, B.J. 2004. The role of Bacillus-based biological control agents in integrated pest management systems: plant diseases. *Phytopathology,* **94**: 1272–1275.

- Jensen, A.B. and Eilenberg, J., 2001. Genetic variation within the insect-pathogenic genus Entomophthora, focusing on the *E. muscae* complex, using PCR-RFLP of ITS II and LSU rDNA. *Mycological Research,* **105**: 307–312.

- Kumar, K., Sridhar, J., Baskaran, R.K.M., Nathan,S.S., Kaushal, P., Dara, S.K. and Arthurs, S., 2019. Microbial biopesticides for insect pest management in India: Current status and future prospects. *Journal of Invertebrate Pathology,* **165**: 74–81.

- Kumar. B.L. and Gopal, V.R.S., 2015. Effective role of indigenous microorganisms for sustainable environment. *Biotech.* **5(6)**: 867–876.

- Mailer, R.J., Scarth, R., Fristensky, B., 1994. Discrimination among cultivars of rapeseed (*Brassica napus* L.) using DNA polymorphisms amplified from arbitrary primers. *Theoretical and Applied Genetics,* **87**: 697-704.

- Mazzola, M., and Freilich, S., 2017. Prospects for biological soil-borne disease control: application of indigenous versus synthetic microbiomes. *Phytopathology,* **107**: 256-263.

- Mishra, J., Tewari, S., Singh, S. and Arora, N. K., 2015. Biopesticides: Where we stand? In -N. K. Arora (eds.), *Plant microbes symbiosis: Applied facets.* New Delhi: Springer, p. 37–75.

- Nielsen, C., Sommer, C., Eilenberg, J., Hansen, K. and Humber, R., 2001. Characterisation of aphid pathogenic species in the genus *Pandora* by PCR techniques and digital image analysis. *Mycologia* **93**: 864–874.

• Ongena, M., Jourdan, E., Adam, A. and Thonnart, P., 2007. Surfactin and fengycin lipopeptides of Bacillus subtilis as elicitors of induced systemic resistance in plants. *Environmental Microbiology*, **9**: 1084–1090.

• Paavanem- Huhtala, S., Avikainem, H., Ylimattila, T., 2000. Development of strain specific primers for a strain of *Gliocladium catenulatum* used in biological control. *European Journal of Plant Pathology*, **106**: 187-198.

• Paran, I., Michelmore, R.W., 1993. Development of reliable PCR-based markers linked to downy mildew resistance genes in lettuce. *Theoretical and Applied Genetics*, **85**: 986-993.

• Peypoux, F., Bonmatin, J.M. and Wallach, J, 1999. Recent trends in the biochemistry of surfactin. *Applied Microbiology and Biotechnology*, **51**: 553–563.

• Ramanujam, B., Rangeshwaran, R., Sivakumar, G., Mohan, M. and Yandigeri, M.S., 2014. Management of insect pests by microorganisms. *Proceedings of Indian National Science Academy*, **80**: 455–471.

• Rohel, E., Couteadier, Y., Paperiok, B., Cavelier, N. and Dedryver, C. A., 1997. Ribosomal internal transcribed spacer size variation correlated with RAPD- PCR pattern polymorphisms in the entomopathogenic fungus *Erynia neoaphidis* isolates. *Mycological Research*, **104**: 213–219.

• Samada, L.H. and Tambunan, U.S.F. 2020. Biopesticides as Promising Alternatives to Chemical Pesticides: A Review of Their Current and Future Status. *Lukmanul Hakim Samada and Usman Sumo Friend Tambunan / OnLine Journal of Biological Sciences* **20 (2)**: 66-76.

• Shahid, M., Srivastava, M., Kumar, V., Singh, A., Pandey, S., 2014. Genetic Determination of Potential *Trichoderma* Species Using ISSR (Microsatellite) Marker in Uttar Pradesh, India. *Journal of Microbial and Biochemical Technology*, **6**: 174-178. doi:10.4172/1948-5948.1000139.

• Sierotzki, H., Camastral, F., Shah, P. A., Aebi, M., and Tuor, U., 2000. Biological characteristics of selected *Erynia neoaphidis* isolates. *Mycological Research*, **104**: 213–219.

• Sobral, B.W. and Honeycutt, R.J., 1993. High output genetic mapping of polyploids using PCR-generated markers. *Theoretical and Applied Genetics*, **86**: 105-112.

- Spurlock, T.N., Rothrock, C.S., Monfort, W.S. and Griffin, T.W., 2016. The distribution and colonization of soybean by *Rhizoctonia solani* AG11 in fields rotated with rice. Soil Biology and Biochemistry, **94**: 29-36.

- Tozlu, E., Tekiner, N., Kotan, R., 2018. Screening of *Trichoderma harzianum rifai*, isolates of domestic plant origin against different fungal plant pathogens for use as biopesticides. *Fresenius Environmental Bulletin*, **27**: 4232-4238.

- Venkatesan, T. and Pattar, S., 2017. Intellectual property guidelines and commercialization of biocontrol agents/biopesticides technologies. In: Varshney*et al.* (Eds.), Training manual on "Identification, Mass Production and Utilization of Parasitoids, Predators and Entomopathogens for Sustainable Insect Pest Management". ICAR-NBAIR, Bengaluru, p. 115–126.

- Welsh, J. and McClelland, M., 1990. Fingerprinting genomes using PCR with arbitrary primers. *Nucleic Acids Research*, **18**: 7213-7218.

- Williams, J.G., Kubelik, A.R., Livak, K.J., Rafalski, J.A. and Tingey, S.V., 1990. DNA polymorphisms amplified by arbitrary primers are useful as genetic markers. *Nucleic Acids Research*, **18**: 6531-6535.

- Yang, X. and Quiros, C.F., 1995. Construction of a genetic linkage map in celery using DNA-based markers. *Genome*, **38**: 36-44.

5

Formulations and Delivery Systems
of *Trichoderma*

S. Sriram and B. Ramanujam
ICAR-Indian Institute of Horticultural Research, Bangalore

Development of Formulations of *Trichoderma* spp.

A formulated product for agricultural application should posses several desirable characters. These include adequate market potential, ease in preparation and application, stability during transportation and storage, abundant viable propagules with good shelf life, sustained efficacy and accepted cost (Churchill, 1982).

Formulation of Antagonistic Bacteria

Talc is usually used as carrier, its pH is adjusted to 7.0 using calcium carbonate. 10g of carboxy methyl cellulose (CMC) per kg of carrier are used as adherent. The inoculum culture of the antagonist (containing a minimum population of 9×10^8 cfu/ml) is mixed with the sterile carrier (400 ml/kg) and air dried. Seeds are coated with a thin layer of 1% CMC and mixed with the respective formulation/s at the rate of 4 g/kg seed.

Shelf life of *Trichoderma* formulation

Shelf life of a bio-control agent plays a crucial role in storing a formulation. In general the antagonists multiplied in an organic food base has greater shelf life than the inert or inorganic food bases. Shelf life of *Trichoderma* in coffee husk was more than

18 months. Talc, peat, lignite and kaolin based formulation of *Trichoderma*, had a shelf life of 4 months. Shelf life of the same in gypsum-based formulation was four months. Studies on the storage of *Trichoderma viride* formulation in poly propylene bags of various colours revealed that the population of *T. viride* was maximum in milky white bags of 100 gauge thickness.

Development of mixed formulations of biocontrol agents

It is likely that most cases of naturally occurring biological control results from mixtures of antagonists, rather than from high population of a single antagonist. For example, mixtures of antagonists are considered to account for protection in disease suppressive soils. Combinations of biocontrol agents for plant diseases include mixtures of fungi and mixtures of fungi and bacteria. Most of these reports on mixtures of biocontrol agents showed that combining antagonists resulted in improved biocontrol. However, there also are reports of combinations of biocontrol agents that do not result in improved suppression of disease compared with the separate antagonists. Incompatibility of the co-inoculants can arise because biocontrol agents may also inhibit each other as well as the target pathogen or pathogens. Thus, an important prerequisite for successful development of strain mixtures appears to be the compatibility of the co-inoculated microorganisms. *Trichoderma harzianum* and *Pseudomomas fluorescens* exhibit better efficacy against soil borne plant pathogens in acidic and neutral to alkaline soils, respectively.

In a study at G.B. Pant Uni. Agri. & Tech, Pantnagar, compatibility of 12 isolates of *T. harzianum* was tested against 41 strains of *P. fluorescens* under *in vitro* condition. In general *P. fluorescens* suppressed growth and sporulation of *T. harzianum* but 4 neutral combinations were identified. One of these combinations TH strain PBAT-43 and PsF strain PBAP-27 were used to develop mixed formulation (Pant Biocontrol Agent-3), which was equally or more effective than individual formulations both under green house and field conditions.

Experiments were conducted at IISR, Calicut to study the population dynamics of fungal and bacterial antagonists in co-inoculated liquid medium and soil, using plate assay for a week. The two-biocontrol agents were found to be compatible with each other and successfully colonized black pepper rhizosphere. A field trial with consortium of biocontrol agents (*Trichoderma aureoviride, Trichoderma harzianum, Trichoderma virens, P. fluorescens* (IISR-6), *P. fluorescens* (IISR-11) was conducted in Kerala against foot rot of black pepper. The best treatment was found to be a

combination treatment of *Trichoderma* spp (Is. No. IISR-143and IISR-369) and
P.fluorescens (IISR-6) in suppression of disease besides increas ing the yield.

Studies at TNAU, Coimbatore showed that *Trichoderma viride* was compatible
with *P. fluorescens*. Seed treatment of chilli with *T. viride* (4g/Kg) and *P. fluorescens*
(5g/Kg) increased germination percentage, shoot length root length and biomatter
production. Vigour index of the seedlings was also increased compared to the
individual applications of the antagonist and control. Combined application of
P.fluorescens and *T.viride* (MNT7) as seed treatment decreased club root of cabbage
and cauliflower under field conditions besides increasing the yield. Combined
application performed better than that of the individual applications.

Table 1. Commercial formulations of biocontrol agents available in India

Product	Bio-agent(s)	Developing agency
Antagon-TV	*T. viride*	Green Tech Agro Propducts, Coimbatore
Biocon	*T. viride*	Tocklai Experimental Station, Tea Research Association, Jorhat, Assam
Bioderma	*T. viride* + *T. harzianum*	Biotech International Limited, New Delhi
Bioguard	*T. viride*	Krishi Rasayan Export Pvt. Ltd., Solan(H.P.)
Bioshield	*Pseudomonas fluorescens*	Anu Biotech International Ltd., Faridabad
Biotok	*Bacillus subtilis*	Tocklai Experimental Station, Tea Research Association, Jorhat, Assam
Ecoderma	*T. viride* + *T. harzianum*	Margo Biocontrol Pvt. Ltd., Bangalore
Ecofit	*Trichoderma viride*	Hoechst and Schering AgrEvo Ltd., Mumbai
Funginil	*T. viride*	Crop Health Bioproduct Research Centre, Gaziabad

Kalisena SD Kalisena SL	*Aspergillus niger AN-27*	Cadila Pharmaceuticals Limited, Ahmedabad
Pant Biocontrol Agent-1	*T. harzianum*	G. B. Pant University of Agriculture Technology, Pantnagar
Pant Biocontrol Agent-2	*Pseudomonas fluorescens*	G. B. Pant University of Agriculture Technology, Pantnagar
Sun–Derma	*T. viride*	Sun Agro Chemicals, Chennai
Trichoguard	*T. viride*	Anu Biotech International Ltd., Faridabad
Tricho-X	*T. viride*	Excel Industries Limited, Mumbai

Table 2: Carrier/food base materials used in *Trichoderma* formulations

Carrier

Pyrax

Celatom, Vermiculite

Vermiculite and Wheat bran

Talc

Diatomaceous earth-molasses

Talc, Gypsum, Lignite, Kaolin, Peat

Wheat bran, Kaolin

Expanded clay, Perlite, Vermiculite

Sodium alginate and CaCl$_2$

Talc, Gypsum, Kaolin, Bentonite, Wheat bran
Rice flour, Gluten, Pyrax, Vermiculite

Production Methods

In general, product formed from solid or semi solid-state fermentation does not require sophisticated formulation procedures prior to use. For example, grain or other types of organic matter upon which antagonists are grown are simply dried ground and added to the area to be treated. There are several problems with solid-state fermentation, which may make the system inappropriate for commercial product development. The preparations are bulky, they may be subject to a greater risk of contamination and they may require extensive space for processing, incubation and storage. The liquid state fermentation is devoid of such problems and large quantities of biomass can be produced within few days. Biomass either can be separated from medium and concentrated or entire biomass with medium can be incorporated into dusts, granules, pellets, wettable powders or emulsifiable liquids. *Trichoderma* spp. can be formulated as pellets (Papavizas and Lewis, 1989), dusts and powders (Luchmeah and Cooke, 1985; Nelson and Powelson, 1988) and fluid drill gels (Conway, et al., 1996; Mihuta and Rowe, 1986). The carrier material may be inert or a food base or a combination of both.

Methodology for preparation of various formulations of *Trichoderma* spp.

1. Talc based formulation

Ingredients:

Trichoderma Culture biomass along with medium	1 liter
Talc (300 mesh, white colour)	2 kg
Carboxymethyl cellulose (CMC) or arabic gum powder	10 kg

Preparation methodology:

Trichoderma is grown in any of the above liquid media for obtaining biomass. After full growth of *Trichoderma*, it is mixed with talc powder in the ratio of 1:2 and dried to 8% moisture under shade. CMC/ Arabic gum powder should be added during mixing. After drying and breaking the clots, the formulation is packed in milky white polythene bags.

2. Vermiculite-wheat bran formulation (Lewis *et al.*, 1991)

Ingredients:

Vermiculite	100 g
Wheat bran	33 g
Wet fermentor biomass	20 g
0.05N HCL	175 ml

Preparation methodology:

Trichoderma is multiplied in molasses-yeast medium for 10 days. Vermiculite and wheat bran are sterilized in an oven at 70 °C for 3 days. Then 20 gms of fermentor biomass and 0.05N HCl are added, mixed well and dried in shade.

3. Pesta granules (Connick *et al.*, 1991)

a. Wheat bran based

Ingredients:

Wheat flour 100g

Fermentor biomass (FB) 52 ml

Sterile water sufficient enough to form a dough

Preparation methodology:

52 ml of FB is added to wheat flour (100g) and mixed by gloved hands to form cohesive dough. The dough is kneaded, pressed flat and folded by hand several times. Then 1 mm thick sheets (pesta) is prepared and air-dried till it breaks crisply. After drying, dough sheet was ground and passed through a 18 mesh (1.0 mm) sieve and granules were collected.

b. Wheat flour- kaolin

Ingredients :

Wheat flour 80 gm

Kaolin 20 gm

Fermentor biomass 52 ml

Preparation methodology: 52 ml of FB is added to wheat flour (100g) and mixed by gloved hands to form a cohesive dough. The rest of the procedure is as described for pesta granules.

c. Wheatflour-bentonite

Ingredients :

Wheat flour 80 gm

Bentonite 20 gm

Fermentor biomass 52 ml

Preparation methodology:

2 ml of FB is added to wheat flour (100g) and mixed by gloved hands to form cohesive dough. The rest of the procedure is as described for pesta granules.

4. Alginate prills (Fravel *et al.*, 1985)

Ingredients:

Sodium Alginate 25 gm

Wheat flour 50 gm

Fermentor biomass 200 ml

Preparation methodology:

Sodium alginate is dissolved in one portion of distilled water (25g/750 ml) and food base is suspended in another portion (50g/250ml). These preparations are autoclaved and when cool are blended together with biomass. The mixture is added drop wise into $CaCl_2$ solution to form spherical beads, which are air-dried and stored at 5 °C.

Delivery of Biocontrol Agents used against Plant Pathogens

Seed treatment

Seed priming, in which seeds are mixed with an organic carrier and then moisture content is brought to a level just below that required for seed treatment which has been used to deliver *T. harzianum* to control *Pythium*- induced damping-off on cucumber. In another process of seed treatment, an industrial film-coating process which was developed for the application of chemicals and biological crop protection agents is being utilized for application of *Trichoderma* spp. *Trichoderma* spp. applied on radish and cucumber seeds through a film coating were shown to be effective against damping-off (Cliquet and Scheffer, 1996).

Soil amendment

Numerous attempts have been made to control several soil borne pathogens by incorporating natural substrates colonized by antagonists of pathogen into soil (Sesan and Csep, 1993). Some examples of diseases controlled by soil amendment of *Trichoderma* spp. are given in Table 2.

Soil drench

Though drenching of soil with aqueous suspensions of bioagents propagules carried out, there will not be any even distribution of bioagents in the soil. It is reported that soil drenching with spore suspension of *T. viride* was very effective in reducing infection from *Colletotrichum truncatum* (brown blotch) infected cowpea seeds. Soil drenching with *T. harzianumm* has given good control of stem rot of groundnut caused by *S. rolfsii* (Kulkarni *et al.*, 1994). An aqueous drench containing conidia of *T. harzianum* controlled wilt of chrysanthemum by preventing reinvasion by *F. oxysporum*.

Aerial spraying / Wound dressing

Trichoderma species can be applied as foliar spray to control diseases affecting above ground parts. Biological control of foliar diseases is not so developed as biocontrol of soil borne diseases. The reasons for the paucity of examples of biocontrol of foliar diseases may be the availability of cheap and effective chemical fungicides and that ease of application to the foliage, and results obtained with biocontrol agents were not so good as those obtained with common fungicides. A more successful example of *Trichoderma* application to aerial plant parts is the biocontrol of wounds on shrubs and trees applied at pruning, in advance of decay fungi (Papavizas, 1985). Grosclaude *et al.* (1970) demonstrated the effectiveness of *T. viride* against *Stereum purpureum*, the cause of silver leaf disease on plum.

Fluid Drill Technology

This delivery system involves the incorporation of biocontrol agents into fluid drill gels. In one study, vegetable or fruit tree seedlings were dipped into gels incorporated with antagonists so that the root area was surrounded by a thin layer of gel before the seedlings were planted. Fluid-drilling gels have been used to deliver *T. harzianum* for control of *R. solani* and *S. rolfsii* on apple (Conaway,1986). This innovative approach, utilizing the benefits derived from fluid drill technology offers considerable promise for the formulation and application of biocontrol microorganisms.

Delivery of Bacterial Antagonists

Four methods viz. a) Root Dip, b) Seed Treatment and c) Soil Treatment are generally followed.

Root dip method

Prepare shake cultures of test bacteria in TSA broth (24 - 48 h). Estimate the population of the cultures by serial dilution.

Seed treatment

Prepare inoculum cultures of antagonist/s and target pathogen as described above and make formulations of the antagonists for seed treatment.

Aerial spray

Biological control of foliar pathogens using bacterial antagonists will necessarily depend upon the establishment and survival of the antagonists on the leaf surface, which has a competitive environment. Addition of nutrients like yeast extract or glucose (jaggery solution) into the inoculum spray has shown to enhance the survival of the antagonists. The talc based formulation is at the rate of 1 Kg/ha as foliar spray.

Use of Talc Formulations of *Trichoderma* for Disease Control

The talc formulation of *Trichoderma* is used for

i.   Seed treatment

ii. Seedling dip

iii. Soil application through farmyard manure for control of diseases like, root rots, wilts, nursery diseases of various crops.

Technique of seed treatment with talc formulation of Trichoderma

Materials required:

»   Talc formulation of *Trichoderma*,

»   Seeds of pulses or vegetables, cereals

»   Small tin

Technique: Seed treatment with talc formulation of *Trichoderma* is done @ 5-10g/ kg seed material depending on the size of the seed. In a small tin, seeds were taken and a lit bit of water is sprinkled on the seed. One teaspoon full (5g) of talc formulation of *Trichoderma* is added to the 1kg of seeds of vegetables/pulse crops and thoroughly mixed to get a fine coating of *Trichoderma* formulation on the seed. Seed treatment should be done just before sowing.

Technique of seedling dip of vegetable and plantation crops in the talc based formulation of *Trichoderma*

Materials required:

»   Talc formulation of *Trichoderma*,

»   Seedlings of vegetable and plantation crops

»   Small tin

Technique:

In a small tin, 10 g (two tea spoons full) of talc formulation of *Trichoderma* is taken and one liter of water is added to it. The suspension is mixed thoroughly and used for seedling dips. Seedlings of tomato, brinjal, chillies, cabbage, beans carrot, etc. which are ready for transplantation are taken and dipped in Trichoderma suspension for a minute. About one liter of the *Trichoderma* suspension may be used for dipping the seedlings required for one acre. Seedling dip should be done just before transplantation. Similarly the seed potato bits and banana suckers can be dipped in the *Trichoderma* suspension for disease control.

Technique of enrichment of Farmyard Manure (FYM) with *Trichoderma* for soil and nursery bed application

Materials required:

» Talc formulation of *Trichoderma* (1kg formulation required for 100 kg of FYM) or *Trichoderma* culture on Sorghum grain or Silkworm excreta (1kg required for 100 kg of FYM)

» FYM-100kg fully decomposed for enrichment (100kg of enriched FYM can be used for one acre)

» Spade

» Water

» Big size Plastic sheet for covering the FYM heaps during incubation.

Technique: Enrichment of FYM should be done in a shaded place either in the field or near the manure pits. 100kg of fully decomposed FYM is spread out on the ground and a little water is sprinkled over it. One kg of Talc formulation of *Trichoderma* or one kg of *Trichoderma* culture on Sorghum grain or one kg of *Trichoderma* culture on Silkworm excreta was uniformly sprinkled on the FYM and thoroughly mixed with a spade. Then the FYM was made into heaps and covered with a plastic sheet. The heaps were kept for 15-20days with intermittent mixing. After 20 days, the FYM would be completely enriched with *Trichoderma* and can be used in the field or nursery beds. In the main field it can be applied at the plant base or in furrows. It can also be applied to the nursery beds for control of nursery diseases and for getting vigorous seedlings of vegetables and plantation crops.

Use of *Trichoderma* formulation prepared from sorghum grains for disease control

The *Trichoderma* formulation prepared from sorghum grains can be used just like talc based formulation of *Trichoderma* for i) Seed treatment ii) Seedling dip iii) soil application through Farmyard manure for control of diseases like, root rots, wilts, nursery diseases of various crops. The dose and technique is same as that of Talc formulation of *Trichoderma*.

Use of *Trichoderma* grown on rice bran/ other agri. by-products for diseases control

Trichoderma grown on rice bran can be used for enriching the FYM for soil application for diseases control in the main field or in nursery beds. The dose and technique is same as that of Talc formulation of *Trichoderma*.

References

- Churchill, B.W. 1982. Mass production of microorganisms for biological control. Pages: 139-156. In: Biological Control of Weeds with Plant Pathogens. Charudattan, R. and Walker, H.L. (eds.). John Wiley & Sons, New York.

- Cliquet, S. and Scheffer, R.J., 1996. Biological control of damping-off caused by *Pythium ultimum* and*Rhizoctonia solani* using *Trichoderma* spp. applied as industrial film coatings on seeds for Biological control of damping-off. *European J. Pl. Pathology*, **102** (3): 247-255

- Connick, W.J., Daigle, D.J. and Quimby, P.C., 1991. An improved invert emulsion with high water retention for mycoherbicide delivery. *Weed Technol.* **5** : pp. 442–444.

- Conway K.E., Tomasino, S. and Claypool, P.L., 1996. Evaluations of biological and chemical controls for southern blight of apple rootstock in Oklahoma nurseries. Proceedings of the Oklahoma Academy of Science, USA, 76, 9–15

- Grosclaude, C., 1970. Preliminary trials of biological protection of pruning wounds against *Stereum purpureum*. *Annales de Phytopathogie*, **2**: 507-516.

- Kulkarni, S.A., Srikant, K. and Kulkarni, S., 1994. Biological control of *Sclerotium rolfsii* Sacc., a causal agent of stem rot of ground nut. *Karnataka Agric. Sci.*, 7: 365-365.

- Lewis, J.A., Papavizas, G.C. and Lumsden, R.D. 1991. A new formulation system for the application of biocontrol fungi to soil. *Biocontrol science and technology*, **1**:59-59.

- Luchmeah, R.S. and Cooke, R.C., 1985. Pelleting of seed with the antagonist *Pythium oligandrum* for biological control of damping-off. *Plant Pathol.* **34**: 528–531.

- Mihuta, L. and Rowe, R.C., 1986. *Trichoderma* spp. as biological control agents of *Rhizoctonia* damping off of radish in organic soil and comparison of four delivery systems. *Phytopathology*, **76**:306-312

• Nelson, M.E. and Powelson, M.L., 1988. Biological control of grey mold of snapbean by *Trichoderma hamatum*. *Plant Disease*, **72**:727-729.

• Papavizas, G.C. and Lewis, J.A., 1989. Effect of *Gliocladium* and *Trichoderma* on damping off and blight of snapbean caused by *Sclerotium rolfsii* in the greenhouse. *Plant Pathology*, **38**:277-286.

• Papavizas, G.C., 1985. *Trichoderma* and *Gliocladium*: Biology and potential for biological control. *Annual review of Phytopathology*, **23**:23-54.

• Sesan, T. and Csep, N., 1993. Prevention of white rot [*Sclerotinia sclerotiorum* (Lib.) de Bary] in sunflower and annual legumes using the biological agent *Coniothyrum* minitans Campbell. *Studii sicercetari de Biologie, Biologie Vegetala* (1991) **43** (1-2): 11-17.

6

Extension of Shelf Life in Formulations of Fungal Bioagents

S. Sriram and B. Ramanujam
ICAR-Indian Institute of Horticultural Research, Bangalore

Introduction

Many commercial products of *Trichoderma* spp. are available for the management of soil borne plant diseases (Fravel, 2005). Among the different propagules like conidia, chlamydospores and vegetative mycelium, the conidia have been the most widely employed in biocontrol programmes (Elad *et al.,* 1993). Conidial biomass can be obtained either by submerged (Elad and Krishner, 1993) or solid substrate (Lewis and Papavizas, 1983) cultivation techniques. Conidia of *Trichoderma* derived from solid state fermentation (SSF) are highly tolerant to abiotic stresses compared with propagules or biomass derived from liquid fermentation (Watanabe *et al.,* 2006). Hence, the formulations derived from liquid fermentation based mass production methods have relatively shorter shelf-life. Though high quality spores could be easily produced by SSF at bench scale, the engineering aspects related to scaling up of SSF have not been developed and applied in many developing countries for mechanized or automated mass production of *Trichoderma*. Wettable powder formulations based on liquid fermentation are popular in India though, they have shorter shelf-life (Singh *et al.,* 2006). Liquid fermentation has the advantages of control over the production process, short time for production, requirement of less space and labour, control over contamination level etc. (Deshpande, 2005). Liquid fermentation can facilitate abundant production of conidial biomass at a shorter period (Harman *et al.*, 1991; Jin *et al.*, 1991). Hence, there is need to extend the shelf life of *Trichoderma* formulations derived from liquid fermentation. Many factors like media and type of inoculum (Elzein *et al.*, 2004), method of drying, addition of protectants (Friesen *et al.*, 2006) and environmental conditions during storage (Connick *et al.*, 1996) affect the viability of the formulation derived from liquid fermentation.

Different interventions can be made during fermentation or at post-fermentation stages to extend the shelf-life of formulations. For example, the addition of chitin as specific nutrient in production medium or in formulation can increase the shelf life of liquid fermentation derived *T. harzianum* talc formulation by two months compared with un-amended formulation or medium. Similarly, heat shock at the end of log phase at 40° C for 30 secs also induced desiccation tolerance and extended the shelf life by 4 months. The osmoticum of the production medium can be adjusted by the addition of poly-ethylene glycol (PEG) or glycerol that can induce trehalose production and provide the desiccation tolerance. Jin *et al.* (1991, 1992, 1996), have shown that desiccation tolerant conidia could be produced by the addition of glycerol in the production medium. Since the biomass of a microbial biocontrol agent is dried after mixing with the carrier material to avoid possible contamination during shelf-life, the conidial biomass has to be desiccation tolerant besides having high spore viability. Compared with PEG, the addition of glycerol was found to be more beneficial since addition of PEG resulted in reduced biomass though it provided desiccation tolerance. Use of glycerol as the osmoticant is particularly effective in initiating micro-cycle conidiation. Accumulation of trehalose that is responsible for stabilizing membranes of cells during desiccation in conidia of *T. harzianum* was correlated with desiccation tolerance (Jin *et al.*, 1996). The shelf life of microbial formulation is affected by the nutrient type in the formulation, moisture content or water activity, osmotic regulation and tolerance of the propagules for adverse environment.

Effect of bioagent-specific nutrients (like chitin and chitosan) in the production media and formulations on the viability and shelf life of *Trichoderma* sp.

Extra nutrients are required in *Trichoderma* formulations for the conidia to germinate, proliferate and become established quickly. An ideal nutrient addition would be available only to the antagonist. The formulation should be designed in such a way that it will not enhance the undesirable organisms to grow. Chitin and chitosan are nutrients degraded by *Trichoderma* but much less so by a range of fungal plant pathogens / contaminants.

Pure chitin (commercial grade) was added to the talc formulations of *T. harzianum* and *T. viride*. Both *T. harzianum* (Th-10) and *T. viride* (Tv-23) were grown on Molasses Yeast Extract Medium in shake culture for 7 days. The talc formulations were prepared and chitin was added at 0, 1, 2, 5 percent level and the formulations were subjected to shelf life studies. During the shelf life studies, it was observed that addition of chitin (2 or 5%) resulted in higher CFU (2×10^6) in the *T. harzianum* formulation upto 6 months while in the control where no chitin was added, the CFU went down below the recommended level of 2×10^6 after 4 months. In chitin amended formulations also, by 7th month the CFU was completely reduced. The chitin addition helped in extending the shelf life by 2 months. In *T. viride* also, the chitin addition helped in extending the shelf life by 2 months.

The colloidal chitin was prepared from pure chitin and added to the formulations of *T. viride* and *T. harzianum* grown in Molasses and yeast extract medium at 0, 0.1, 0.2 and 0.5% concentrations. The formulations were tested for shelf life. Addition of colloidal chitin at 0.1 and 0.2 percent levels helped in maintaining the high CFU in the formulations of *T. viride*. In higher concentration (0.5%) after 5 months, there were other fungal contaminations leading to the reduced population of *T. viride*. Addition of colloidal chitin helped in extending the shelf life by 2-3 months compared to control. In formulations were chitin was not added, the CFC could not be detected after 5th month of storage. In *T. harzianum* also, the shelf life is extended by two months.

Effect of Carbon:Nitrogen ratio of production media on spore longevity

C:N ratio and non carbon concentration appears to play a key role in spore longevity (Agosin and Aguilera, 1998). At pH 7.0, a medium having C:N of 14:1 consistently produced conidia with longer shelf life because in this medium, both C and N are consumed simultaneously and once they are depleted, the sporulation starts along with the onset of synthesis of other metabolites involved in spore survival. To study the effect of C:N ratio on the shelf life, *T. viride* (Tv-23) and *T. harzianum* (Th-10) were grown on the *Trichoderma* Specific Medium (TSM) with glucose as carbon source and ammonium nitrate or cassamino acids as N source. In both *T. viride* and *T. harzianum,* the C:N ratios, 1:5 and 1:1 did not promote the growth as well as the shelf life of both *T. viride* and *T. harzianum.* The wider the ratio, higher was the CFU and in shelf life also, 10:1 and 15:1 ratios helped in having higher CFU upto 6 months only. Since in this experiment, only synthetic medium was used, the CFU level was also affected, as there was no rich food base in the formulation. The higher C:N ratio is already being reflected in the composition of media that was used for mass production (Jaggery or Molasses – 30g per litre and Yeast extract or Soya powder – 10g per litre) which gave high propagules per unit volume.

Effect of addition of osmoticants/humectants to the production media on the induction of desiccation tolerance in the propagules of *Trichoderma* sp.

To alleviate the problem of conidia drying, appropriate osmoticants (like poly ethylene glycol or glycerol) was included in the medium up to 2 Mpa in the medium of production. Timing of addition of osmoticum is important. With the addition of osmoticum, hyphal production decreased with conidia production being enhanced (Jin *et al.*, 1991, 1996). PEG 8000 or PEG 200 can be used to reduce the viscosity problems and to increase oxygen transfer rate PEG 200 can be used. PEG 200 is less costly than PEG8000. Glycerol at 9%, is also appropriate and cost effective. Addition of glycerol in production medium at 3% concentration increased the retention of viability of propagules upto 8 months (10^{11}) compared to control where glycerol was not added (10^6 by 6th month). Addition of glycerol at 6 and 9% also increased the viability but lower than that by 3% glycerol addition. Though water activity got reduced in all formulations after 6 months, in treatments with 6 and 9% glycerol addition, the viability got reduced significantly while propagules in treatment with 3% addition, were tolerant up to 12th month.

Effect of induction of trehalose accumulation on tolerance of propagules to adverse environments

Trehalose is a reserve carbohydrate found in fungal spores. It serves as stress protectant also. In the stationary sporulation stage, the trehalose production is higher than that of exponential stage of sporulation. Heat shock at the end of the sporulation phase resulted in the production of heat-resistant conidial biomass with high shelf life. Heat shock at 40° C for 90 min induces trehalose accumulation (Pedreschi *et al.*, 1997). Trehaloses are produced at the time of stress. If trehalose accumulation is induced it will help the propagules to have good shelf life. Induction of trehalose accumulation was done by heat shock method (32 to 42° C) at the stationary sporulation stage and its effect on spore longevity during storage was studied for *Trichoderma* sp. The heat shock at the end of log phase of fermentation (just before harvest) induces the stress in culture and during this stationary phase with stress, there is increased production of trehalose in fungi that helps in desiccation tolerance. To study the effect of heat shock, *T. harzianum* culture was grown in 10 litre fermentor and subjected to heat shock at 40° C for 30 min or 40° C for 60 min at the end of log phase of the fermentation (i.e. 72 h). One treatment without heat shock was maintained as control. The talc formulations were packed at 15% moisture and then packed in polypropylene bags. They were stored at room temperature. The moisture content, water activity and viability of propagules were recorded at monthly interval.

Effect of residual moisture content/water activity/equilibrium relative humidity in the production medium and formulation

Water activity (a_w) is the ratio between the partial pressure of water in a material and the vapour pressure of pure water at the same temperature under equilibrium conditions. Silica gel can give 0.03 (a_w). Water activity principles have been applied to the formulation of microorganisms intended for agricultural use, such as bacteria (Mugnier and Jung, 1985), fungi and nematodes (Connick *et al.*, 1993, 1994). Studying water activity is more meaningful than water or moisture content in case of dried preparations of microorganisms. A low a_w often increases the resistance of a fungus to high temperature and favours long-term viability. 15% initial moisture content was found to help in retaining more viable propagules for longer period (by additional one month) compared to storing with initial moisture content of 8 or 10%. Though storing at 20% moisture content helped in retaining more propagules, the contamination levels were high. More than the moisture content, water activity was found to be crucial. Three different packing methods *viz.* normal packing, vacuum packing and nitrogen packing did not differ significantly with respect to viability of propagules when formulations were stored with different initial moisture levels. Once the water activity came down drastically, the viability also came down. The packing material that holds the water activity has to be used.

Effect of combination of different treatments like addition of chitin and glycerol with heat shock on the shelf life of *T. harzianum*

Twelve combinations were made with treatments that were found to be effective in preliminary experiments *viz.*, addition of colloidal chitin in production medium, addition of glycerol at 3 or 6% and heat shock. Addition of glycerol at 3% concentration helped in extending the shelf life upto 10 months with or without heat shock. In control, the CFU recovery was less than 10^6 after 7th month. Similarly, addition of glycerol along with colloidal chitin at 0.2% also resulted in shelf life of 9-10 months. Addition of glycerol at 6% resulted in shelf life of 15-16 months with or without heat shock. Addition of colloidal chitin alone at 0.2% did not significantly change the CFU recovery after 4 months. However, the initial population was high when 0.2% colloidal chitin was added in the fermentation media. Addition of 0.2% colloidal chitin along with 6% glycerol resulted in prolonged shelf life compared to addition of glycerol alone at 6% concentration. The CFU recovery was above 10^6 even at 17th month in these formulations derived from liquid medium amended with 6% glycerol and 0.2% colloidal chitin with or without heat shock.

In bio-efficacy tests, with the treatment of formulations derived with different interventions either alone or in combination, there was significant reduction in wilt disease incidence compared to pathogen inoculation alone. In pathogen alone inoculated plants, the wilt incidence was 72% while in treated plants it varied from 0 to 30.59%. At the end of 15 months of shelf life, the plants could be protected from wilt incidence by these formulations.

Effect of different packing materials on the shelf life of *T. harzianum* talc formulation

Different packing materials were used and the most vulnerable ones were foam type, BST (75μ), polypropylene and HMHDP (25μ). Shelf life of 3 months only could be achieved with these packing materials. In aluminium foil covers, the shelf life could be extended upto 4 months while in LLDP 80μ, the shelf life could be extended upto 5 months with the talc formulations without any other treatments to enhance shelf life.

Solid substrate fermentation based talc formulation

Solid substrate-derived conidial formulations with or without substrate incorporation were found to have more shelf life than formulations prepared using biomass from liquid fermentation. The viability of propagules at 10^8 cfu/g could be obtained in the SSF derived formulations even after 8th month. They were found to be tolerant to low water activity. Different drying methods were used before packing. The shelf life of these formulations were 14 months with recovery of CFU above 10^6 in normal packing while in vacuum packing it was 12 months. In the bioefficacy studies also, the products that were processed with drying methods were able to reduce wilt incidence in tomato upto 35% compared to control which was 74%.

Effect of methods of drying on shelf life of *T. harzianum* derived from SSF

Conidia based formulations derived from SSF were prepared and subjected to different drying methods. The drying methods used as post production intervention, helped in extending the shelf life of the formulation upto 14-15 months compared to control that had shelf life of 7-8 months. Drying methods were effective in extending the shelf life of *Trichoderma* formulations upto 15 months after which the recovery of CFU per g was less than 10^6. The bio-efficacy of these formulations was tested using *Fusarium lycopersici* as target pathogen and tomato wilt as the host-pathogen system. Drying methods were found to be better with respect to bio-efficacy of the formulations. With the pathogen inoculation alone the percent wilt incidence was 69% while in treated plants it was ranging from 13.11% to 65%.

Effect of forced drying on the shelf life of SSF based talc formulations of *T. Harzianum*

One set of conidia based formulations with an additional step of post-production processing, was studied for shelf life. The forced drying was included as additional step in post production processing. One set of formulations that were subjected to normal drying packed in normal or vacuum conditions was also prepared separately for comparison. The CFU recovery was upto 20 months which was $>10^6$ CFUs g^{-1} in both normally packed formulations subjected to forced drying. In formulations with modified packing process that was subjected to forced drying, the shelf life was only upto 18 months after which the population was less than 10^6 CFU g^{-1}.

References

• Agosin, E. and Aguilera, J.M., 1998. Industrial production of active propagules of *Trichoderma* for agricultural uses. *In*: G.E. Harman and C.P. Kubicek (eds.) *Trichoderma* and *Gliocladium* Vol. 2. Enzymes, Biological Control and Commercial Applications. Taylor and Francis, London, pp. 205-207.

- Connick, W.J. Jr., Nickle, W.R. and Vinyard, B.T., 1993. Pesta: New granular formulations of *Steinernema carpocapsae*. *Journal of Nematology*, **25**: 198-203.

- Connick, W.J. Jr., Nickle, W.R., Williams, K.S. and Vinyard, B.T., 1994. Granular formulations, *Steinernema carpocapsae* (Nematoda: Rhabditida) with improved shelf life. *Journal of Nematology*, **26**: 352-359.

- Connick, W.J. Jr., Daigle, D.J., Boyette, C.D., Williams, K.S., Vinyard, B.T. and Quimby, Jr. P.C., 1996. Water activity and other factors that affect the viability of *Colletotrichum truncatum* conidia in wheat flour-kaolin granules (pesta). *Biocontrol Sci. Tech.*, **6**: 277-284.

- Deshpande, M.V., 2005. Formulations and applications of mycopathogens. *In: Microbial Biopesticides: Formulations and Application*, R.J. Rabindra, S.S. Hussaini and B. Ramanujam, (eds.) Bangalore, India. Project Directorate of Biological Control, 2005, pp.150-158.

- Elad, Y. and Krihsner, B., 1993. Survival in the phylloplane of an introduced biological control agent (*Trichoderma harzianum*) and populations of the plant pathogen *Botrytis cinerea* as modified by abiotic conditions. *Phytoparasitica*. **21**: 303-313.

- Elad, T., Zimand, G., Zaqs, Y., Zuriel, S. and Chet, I., 1993. Use of *Trichoderma harzianum* in combination or alteration with fungicides to control cucumber grey mold (*Botrytis cinerea*) under commercial greenhouse conditions. *Biol. Control.*, **42:** 324-332.

- Elzein, A., Kroshel, J. and Mueller-Stoever, D., 2004. Effects of inoculum type and propagule concentration on shelf life of pesta formulations containing *Fusarium oxysporum* Foxy2, a potential mycoherbicide agent for *Striga* spp. *Biol. Control.*, **30**:203-211.

- Fravel, D.R., 2005. Commercialization and implementation of biocontrol. *Ann. Rev. Phytopathol.*, **43**:337-359.

- Friesen, T.J., Holloway, G., Hill, G.A. and Pugsley, T.S., 2006. Effect of conditions and protectants on the survival of *Penicillium bilaiae* during storage. *Biocontrol Sci. Tech.*, **16** (1):89-98.

- Harman, G.E., Jin, X., Stasz, T.E., Peruzzotti, G., Leopold, A.C. and Taylor,

A.G., 1991. Production of conidial biomass of *Trichoderma harzianum* for biological control. *Biol. Control.*, **1**:23-28.

• Jin, X., Hayes, C. and Harman, G.E., 1992. Principles in the development of biological control systems employing *Trichoderma* species against plant pathogenic fungi. *In*: *Frontiers in Industrial Mycology* (Ed.) G.F. Leatham, Chapman and Hall, New York.

• Jin, X., Taylor, A.G., and Harman, G.E., 1991. Conidial biomass and desiccation tolerance of *Trichoderma harzianum* produced at different water potentials. *Biol. Control.* **1**:237-243.

• Jin, X., Taylor, A.G. and Harman, G.E., 1996. Development of media and automated liquid fermentation methods to produce desiccation tolerant propagules of *Trichoderma harzianum. Biol. Control.* **7**:267-274.

• Lewis, J.A. and Papavizas, G.C., 1983. Chlamydospore formation by *Trichoderma* spp. in natural substrates. *Canadian Journal of Microbiology.* 30: 1-6.

• Mugnier, J. and Jung, G. 1985. Survival of bacteria and fungi in relation to water activity and the solvent properties of water in biopolymer gels. *Applied and Environmental Microbiology.* **50**: 108-114.

• Pedreschi, F., Aguilera, J.M., Agosin, E. and SanMartin, R. 1997. Induction of trehalose in spores of the biocontrol agent, *Trichoderma harzianum. Bioprocess and Biosystem Engineering.* **17**:317-322.

• Singh, U.S., Zaide, N.W., Joshi, D., Vashney, S. and Khan, T., 2006. Current status of *Trichoderma* spp. for the biological control of plant diseases. *In: Microbial Biopesticides: Formulations and application*, R.J. Rabindra, S.S. Hussaini, B. Ramanujam, (eds.) Bangalore, India: Project Directorate of Biological Control, India, pp.13-48.

• Watanabe, S., Kato, H., Kumakura, K., Ishibashi, E. and Nagayama, K., 2006. Properties and biological control activities of aerial and submerged spores in *Trichoderma asperellum* SKT-1. *J Pesti. Sci.*, **31**(4):375-379.

7

Current Scenario on Use of *Pseudomonas* spp. in Horticulture

Ashwitha K.[1] and Rajagopal Rangeshwaran[2]
[1]Centre for Cellular and Molecular Platforms, Bengaluru
[2]National Bureau of Agricultural Insect Resources, Bengaluru

Introduction

Soil is a complex entity harboring a plethora of micro-organisms which are soul for many activities which benefits as well as detriments the soil condition and health of the plants which grow from it. A huge amount of data is available on the use of rhizobacteria for plant growth promotion in natural or artificial ecosystems. The primary criterion of any plant growth promoting bacteria lies is in its root colonization capacity. The colonization of bacteria in different plant niches will provide more knowledge on plant-microbe interactions (Glick, 1995; Picard and Bosco, 2008).

The beneficial microorganisms which are extrapolated for plant health and disease control are termed as microbial biological control agents (MBCAs). *Pseudomonas* is one of the proven biocontrol agents known to improvise the plant health and also controls plant diseases through varied mechanisms (Kumar *et al.*, 2017). Broadly *Pseudomonas* spp. can be termed as a bio product due to its versatile mechanisms. It is an extensively studied bacteria and it has been accepted among the scientific and farming community for its benefits. In the current era, global food production demand has increased however at the same time the use of chemical pesticides and fertilizers is alarming. This situation has brought interest in the farming community to adopt sustainable methods of agricultural practices to maintain soil health. The scientific community has been working to develop potential strains of *Pseudomonas* sp. which can improve plant health, control fungal and bacterial infections in plants, nutrient acquisition and stress mitigation. However, the adoption of the MBCAs has received far less importance than its research. The major reason is its inefficacy to perform in all the soils and climatic conditions. Researchers are focusing to unmask the lack of efficacy in their performance in different soil types, climatic conditions, variety of the crop, bacterial genotype and proper inoculation technology (Baez-Rogelio *et al.*, 2017). Advanced biocontrol research is focusing on selecting a best strain by decoding multi trophic interactions of the microorganisms to overcome this hurdle (Kohl *et al.*, 2019).

Pseudomonas spp. as a potential bio agent?

Rhizosphere is a microbial rich zone and the *Pseudomonas* spp. flourish by using the root exudates as a source of nutrition (Picard *et al.*, 2000). The root exudates of plants are essential for proliferation of bacteria in the rhizosphere. The bacteria can survive even under different stress conditions in soil as they are capable of expressing additional stress tolerance related genes in presence of plant nutrients which were not present in proteome of bacteria alone (Afroz *et al.*, 2013; Calvo *et al.*, 2014). *Pseudomonas* is a gram negative aerobic bacteria that is ubiquitous in soil and grows in abundance in root zones. They comprise of *Pseudomonas aeruginosa*, *P. aureofaciens*, *P. chlororaphis*, *P. putida* (two biotypes), *P. fluorescens* (four biotypes), and plant pathogenic species *P. cichorii* and *P. syringae*. All fluorescent pseudomonads fall into one of the five ribonucleic acid homologies. Specific strains of fluorescent *Pseudomonas* sp. inhabit the environment surrounding plant roots and some even the root interior. They influence the growth of the plant through several mechanisms which can be broadly classified into direct and indirect growth promotion. Direct growth promotion occurs when *Pseudomonas* synthesizes phytohormones, solubilizes minerals and mitigates stress. Indirect growth promotion or biocontrol occurs when they reduce or prevent the deleterious effects of phytopathogenic organisms. This can be achieved by the induction of systemic resistance or by the synthesis of antimicrobial compounds. *Pseudomonas* sp. have been observed to impart salt and drought tolerance in paddy, lettuce, maize, groundnut (Paul and Nair 2008). The abiotic stress mitigation is observed in rhizospheric soils of groundnut, black gram, green gram, red gram, soybean, sunflower, maize and rice soils of Telangana which were treated with high temperature stress tolerant strains of *Pseudomonas* spp. There is no specific scientific evidence for abundance of bacteria with respect to crops. However, *Pseudomonas* spp. is reported in diverse rhizospheric habitats (Mercado-Blanco and Bakker, 2007; Hayat *et al.*, 2010). These studies increase scope of obtaining a suitable isolate for respective pathogens (Charulatha *et al.*, 2013). Introducing these bacterial strains to plant roots can lead to increased plant growth, usually due to suppression of plant pathogenic microorganisms (Mercado-Blanco and Bakker, 2007).

Promotion of *Pseudomonas* strains which can mitigate the stress in high salt conditions, drought inflicted soils can improve the plant health even under adverse conditions (Khare *et al.*, 2011). *P. fluorescens* and *P. putida* produce an array of heat stress and salt stress tolerant genes under extreme conditions which helps it to influence better plant growth. Proteomic studies have revealed that *Pseudomonas* species regulated nutrient uptake, synthesize osmoprotectants under water limiting conditions to favor its survival (Rangeshwaran *et al.*, 2013; Ashwitha *et al.*, 2018). The combination of experiments along with studying its tolerance to abiotic stresses and characterization for plant growth promoting attributes is useful in choosing a versatile strain of plant growth promoting bacteria which can be exploited for commercial value (Walsh *et al.*, 2001). *P. fluorescens* is the majorly commercialized bacterium however, many other species of *Pseudomonas* are potential to be used in promotion of plant health. The current research should enable us to utilize even these strains specific to certain soils and crops for better results.

Benefits of *Pseudomonas* as a Bio-agent in Horticulture Crops

It is well understood that *Pseudomonas* spp. are effective soil dwellers but their population depends on different factors which are bound to change due to biotic and abiotic factors. The major factors which can be listed are pH, salinity, soil type, soil structure, soil moisture, soil organic matter, plant exudates and also other external environmental conditions including climate, pathogen presence etc. In addition, we cannot deny human practices as also one of the major factors in deterioration of soil condition. Even under these circumstances, there are umpteen researches which prove that strains of *Pseudomonas* spp. can grow under extreme conditions. The major role of this bacterium can be classified as discussed below:

1. Biofertilizers

Soil contains minerals in free forms or unavailable to plant roots. *Pseudomonas* has been known to solubilize major limiting nutrients like nitrogen and phosphorus to roots (de Freitas *et al.*, 1997). *Pseudomonas* sp. solubilise the nutrients by production of organic acids. Gluconic acid is produced by *Pseudomonas* sp to convert inaccessible forms of phosphorus to readily absorbable forms to plant roots (Kalayu, 2019). It is even interesting that *Pseudomonas fluorescens* strain can solubilize calcium phosphate at different pH levels. Some of the studies have revealed that *Pseudomonas* sp. have the ability to regulate the expression of enzymes and activate it at low bioavailability of phosphate content. In fact *Pseudomonas* spp. has been known to produce acidic phosphatase expressions in them and enable the plant growth even under high stress conditions.

2. Plant growth stimulator

Pseudomonas spp. has shown to alleviate salt stress, drought stress, nutrient stress, heavy metal toxicity and temperature stress in different crops. The mechanisms to overcome these stress conditions could be accumulation of osmoprotectants, biosynthesis of antioxidative enzymes, through enhancement of nutrient uptake (Ashwitha *et al.*, 2018). Most commonly discussed benefits of *Pseudomonas* are the ability to produce a vital enzyme, (1) 1-aminocyclopropane-1-carboxylate (ACC) deaminase to reduce the level of ethylene in the root of developing plants thereby increasing the root length and growth; (2) the ability to produce hormones like auxin, i.e. indole acetic acid (IAA) (Patten and Glick, 2002), abscisic acid (ABA), gibberellic acid (GA) and cytokinins. Seed bacterization method is effective in enhancing growth and also shows better inhibition to fungal pathogens (Ardebili *et al.*, 2011). Strains like *Pseudomonas extremorientalis* TSAU20 and *P. chlororaphis* TSAU13 have even shown effective colonization in common bean seedlings under saline conditions resulting in increased root and shoot growth compared to control under saline conditions. Production of IAA, siderophore and pyocyanin have even resulted in alleviation of the stress response in chickpea (Khare *et al.*, 2011).

3. Biopesticides

Pseudomonads are under constant interactions with various microbes in soil. They produce antimicrobial metabolites which suppresses major pathogens in soil. *Pseudomonas* spp. has the ability to produce a wide variety of antibiotic compounds which are specific or have a wide application on control of pathogens. The importance of registration of effective strains will add value for commercialization (Walsh *et al.*,

2001). *P. fluorescens* strains Pf-5 and CHAO are identified for its antibacterial activity and ability to improve plant health (Scales *et al.*, 2014). One of the early reports which described the biocontrol traits of *Pseudomonas* sp. involved in suppression of major plant pathogens through scavenging iron in the rhizosphere environment through release of siderophore, production of antimicrobial metabolites like Phenazines, 2,4-diacetylphloroglucinol (DAPG), pyoluteorin, pyrrolnitrin, hydrogen cyanide (HCN), and cyclic lipopeptides. This information was valuable when choosing the right strain as a biological control agent (O'Sullivan and O'Gara, 1992). Pathogen signalling has a major role to play in disease protection by microbes in plants. DAPG producer CHAO chose hydrogen cyanide as biocide to repress the pathogen *Fusarium oxysporum* f.sp. *radicis lycopersici* unlike *P. fluorescens* Q2-87 which showed DAPG as its major biocontrol mechanism against *F. oxysporum*. *Pseudomonas* strain reveals a differential response in multi trophic plant microbe interactions (Duffy *et al.*, 2004). Root exudates have a key role to play in interaction of plant growth promoting bacteria and its expression of biocontrol traits (Dubuis *et al.*, 2007).

Insecticidal activity of *Pseudomonas* is not much tested and tried in the field. However, there are numerous reports of *Pseudomonas* strains like CHAO, Pf-5, F6 and many others which has shown effect on *Galleria mellonella*, *Spodoptera littoralis*, *Drosophila melanogaster* etc through production of toxins, biocide or biosurfactant (Kupferschmied *et al.*, 2013).

Popularization of *Pseudomonas* in horticulture

The commercial exploitation of *Pseudomonas* in agriculture and horticulture is done for more than two decades now. The usage of this bio product can be enhanced only with demand for organic horticulture. Currently, organic practices are not every growers choice as the productivity is less compared to conventional practices (Pylak *et al.*, 2019). *Pseudomonas* can act at different capacities in the rhizosphere, non rhizosphere, phylloplane and endophytically. Hence, *Pseudomonas* spp. demonstrate as biofertilizers, biopesticides, bio stimulants and bioenhancers (Woo and Pepe, 2018). Few of the European countries have started production of *Pseudomonas* as bio stimulants. Increased demand for food production has created markets for chemicals or botanicals which give instant results in the crops (Calvo *et al.*, 2014). In spite of a large scientific force working in promotion of biological control agents or bio stimulants for sustainable agricultural practices, *Pseudomonas* has taken a slow pace in reaching the farming community. As discussed earlier, the major challenge is the performance of a single product across all the soils and climatic conditions. The onus is on the entire scientific community to create awareness to reach even the last mile. In India, major populations of farmers are still unaware of existing technologies developed for the betterment of plant health. Parallely even the business of bio pesticides and biological control products concentrates in developing a single organism having multitude of activities. With advances in technologies, it is imperative that even the regulations regarding pesticides/fertilizers need a revamp. This will enable introduction of newer microbial based products and cell free microbial based products which are more rigorous and effective.

Conclusion

Horticultural farmers are showing interest in shifting to organic cultivation practices. However their major concern is on knowledge dissemination on proper methods, availability on inputs and sources where they can procure the bio products. The natural instinct of a farmer is to attend the problems after they are evident in the crop. An immediate resort to the pest incidence or a pathogen infestation makes a farmer switch to unregulated usage of chemicals. The mundane activity of application of unscrupulous quantities of the chemical fertilisers to benefit the crop for a short period in a continuous cycle has damaged the soil dynamics beyond limits. An increased awareness has to be created on proper application practices and delivery methods for an effective pathogen or pest control for plant health. The use of *Pseudomonas* sp. alone in a conventional farming may not be enough during a cropping cycle but a combined application of *Pseudomonas* with other microbial products, botanical or essential chemical can increase the chances of better plant resistance, stress mitigation and plant growth. The onus is on the scientific researchers, farm extension officers, agricultural officers and the bioproduct promoters to carry out proper awareness among the farming community on use of *Pseudomonas* based products. Different delivery mechanisms of *Pseudomonas* formulation like seed bacterization, farm yard manure treatment, drenching, furrow application and foliar spray can enhance its activities. The most crucial step in adopting these microbial based products is its time of application. Unlike a chemical fertilizer or pesticide, these bioproducts perform efficiently only if applied as prophylactic. Next generation sequencing is bringing a paradigm shift in selection of potential strain which has advantages over traditional techniques of isolation through screening at laboratory level. The advanced technologies will bring clarity on concerns regarding performance of biocontrol agents, effect of application, and survivability of biocontrol agents in soil after application (Jijakli *et al.*, 2015). The research will concentrate on understanding the bi/tri trophic interactions of the selected bioagent.

Application of microbial consortia is one way to enhance the efficacy compared to application of a single strain of bacteria. Studies have shown that *Pseudomonas* consortia could survive under diverse ecosystems and perform better in tackling the pathogen (Compant *et al.*, 2019). It is much more important now to design smart microbial consortia, adopt practices which favour the microbial biota, modified plant breeding approaches, development of suitable formulations and delivery approaches. Microbial based products are environmentally safe approaches, cost effective, easy to produce and also have multiple actions against plant pest/pathogens for sustainable agriculture.

References

- Afroz., A., Zahur., M., Zeeshan., N. and Komatsu., S., 2013. Plant bacterium interactions analysed by proteomics. *Frontiers in Plant Science*, **4**: 1.

- Ardebili, Z.O., Ardebili, N.O. and Hamdi, S.M.M., 2011. Physiological effects of *Pseudomonas fluorescens* CHA0 on tomato (*Lycopersicon esculentum* Mill.) plants and its possible impact on *Fusarium oxysporum* f. sp. Lycopersici. *Australian Journal of Crop Science*, **5**: 1631–1638.

- Ashwitha, K., Rangeshwaran, R. and Sivakumar, G., 2018. Molecular Mechanisms adopted by abiotic stress tolerant Pseudomonas fluorescens (NBAII-PFDWD) in response to in vitro osmotic stress. *Journal of Biological Control*, **32**: 52-61.

- Baez, A., Moraler-Garcia, Y.E., Hernandez, V.C. and Muoz-Rojas, J., 2017. Next generation of microbial inoculants for agriculture and bioremediation. *Microbial Biotechnology*, **10**:19-21.

- Calvo, P., Nelson, L. And Kloepper, J.W. 2014. Agricultural uses of plant biostimulants. *Plant Soil* DOI 10.1007/s11104-014-2131-8.

- Charulatha, R., Harikrishnan, H., Manoharan, P.T. and Shanmugaiah, V., 2013. Characterization of groundnut rhizosphere *Pseudomonas* sp. VSMKU 2013 for control of phytopathogens. *In: Microbiological Research In: Agroecosystem Management*, Springer India. 121-127pp.

- Compant, S., Samad, A., Faist, H. And Sessitsch, A., 2019. A review on the plant microbiome: Ecology, functions, and emerging trends in microbial application. *Journal of Advanced Research,* **19**: 29–37.

- de Freitas, J.R., Banerjee, M.R., Germida, J.J., 1997. Phosphate-solubilizing rhizobacteria enhance the growth and yield but not phosphorus uptake of canola (Brassica napus L.). *Biology and Fertility of Soils,* **24**: 358–364.

- Dubuis, C. and Keel, C., 2007. Dialogues of root-colonizing biocontrol pseudomonads. *European Journal of Plant Pathology,* **119**: 311 – 328.

- Duffy, B., Keel, C. and Defago, G., 2004. Potential role of pathogen signalling in multitrophic plant-microbe interactions involved in disease protection. *Applied and Environmental Microbiology,* **70**: 1836 – 1842.

- Glick, B.R., Karaturov'ic, D.M. and Newell, P.C., 1995. A novel procedure for rapid isolation of plant growth promoting pseudomonads. *Canadian Journal of Microbiology,* **41**:533–536.

- Hayat, R., Ali, S., Amara, U., Khalid, R. and Ahmed, I., 2010. Soil beneficial bacteria and their role in plant growth promotion: a review. *Annals of Microbiology*

- Jijakli, H., Margarita, M.M. and Sébastien, M. 2015. Biological control in the microbiome era: challenges and opportunities. *Biological Control* doi: http://dx.doi.org/10.1016/j.biocontrol.2015.06.003

- Kalayu, G., 2019. Phosphate Solubilizing Microorganisms: Promising Approach as Biofertilizers. *International Journal of Agronomy.* doi.org/10.1155/2019/4917256

- Khare, E., Singh, S., Maheshwari, D.K. and Arora, N.K., 2011. Suppression of charcoal rot of chickpea by fluorescent *Pseudomonas* under saline stress condition. *Current Microbiology* **62:** 1548 – 1553.

- Köhl, J., Kolnaar, R. and Ravensberg, W.J., 2019. Mode of Action of Microbial Biological Control Agents Against Plant Diseases: Relevance Beyond Efficacy. *Frontiers in Plant Science,* **10**:845. doi: 10.3389/fpls.2019.00845.

- Kumar, A., Verma, H., Singh, V.K., Singh, P.P., Singh, S.K., Ansari, W.A., Yadav, A., Singh, P.K. and Pandey, K.D., 2017. Role of *Pseudomonas* sp. in Sustainable Agriculture and Disease Management *In: Agriculturally Important*

Microbes for Sustainable Agriculture. Vol 2. Application in Crop Production and Production. (Ed). Meena, V.S., Mishra, P.K., Bisht, J.K. and Pattanayak, A. Springer doi:10.1007/978-981-10-5343-6_7

• Kupferschmied, P., Maurhofer, M. And Keel, C., 2013. Promise for plant pest control: root-associated pseudomonads with insecticidal activities. *Frontiers in Plant Science*, **4**: doi: 10.3389/fpls.2013.00287

• Mercado-Blanco, J. and Bakker, P.A.H.M., 2007. Interactions between plants and beneficial *Pseudomonas* spp.: exploiting bacterial traits for crop protection. *Antonie van Leeuwenhoek*, **92**: 367 – 389.

• O'Sullivan, D. and O'Gara, F., 1992. Traits of fluorescent *Pseudomonas* spp. involved in suppression of plant root pathogens. *Microbiology Reviews*, **56**: 662 – 674.

• Patten, C.L. and Glick, B.R., 2002. Role of Pseudomonas putida Indoleacetic acid in development of host plant root system. Applied and Environmental Microbiology, 68: 3795-3801.

• Paul, D. and Nair, S., 2008. Stress adaptations in a plant growth promoting rhizobacterium (PGPR) with increasing salinity in the coastal agricultural soils. *Journal of Basic Microbiology*, **48**: 378 -384.

• Picard C., Di Cello F., Ventura M., Fai R. and Guckert, A., 2000. Frequency and biodiversity of 2,4-diacetylphloroglucinolproducing bacteria isolated from the maize rhizosphere at different stages of plant growth. *Applied and Environmental Microbiology*, **66**: 948-955.

• Picard, C and Bosco, M., 2008. Genotypic and phenotypic diversity in populations of plant probiotic *Pseudomonas* spp. colonizing roots. *Naturwissenschaften*, **95**: 1-16.

• Pylak, M., Oszust, K. And Frac, Magdalena., 2019. Review report on the role of bioproducts, biopreparations, biostimulants and microbial inoculants in organic production of fruit. *Reviews on Environment, Science and Biotechnology*, **18**:597–616.

• Rangeshwaran, R., Ashwitha, K., Sivakumar, G. and Jalali, S.K., 2013. Analysis of proteins expressed by an abiotic stress tolerant Pseudomonas putida (NBAII-RPF9) isolate under saline and high temperature conditions. *Current Microbiology*, **67**:659-667.

- Scales, B.S., Dickson, R.P., LiPumac, J.L. and Huffnaglea, G.B., 2014. Microbiology, Genomics, and Clinical Significance of the *Pseudomonas fluorescens* Species Complex, an Unappreciated Colonizer of Humans. *Clinical Microbiology Reviews*, **27**: 927–948.

- Walsh, U.F., Morrissey, J.P. and O'Gara, F., 2001. *Pseudomonas* for biocontrol of phytopathogens: from functional genomics to commercial exploitation. *Current Opinion in Biotechnology*, **12**: 289 – 295.

- Woo SL and Pepe O., 2018. Microbial Consortia: Promising Probiotics as Plant Biostimulants for Sustainable Agriculture. *Frontiers in Plant Science*, **9**:1801. doi: 10.3389/fpls.2018.01801.

8

Bio-Formulations as a Component of IDM of Blight and Wilt in Pomegranate

G. Manjunath

University of Horticultural Sciences, Bagalkot

Introduction

Pomegranate (*Punica granatum* L.) is an important fruit crop of subtropical and tropical regions of the world. It is promoted as functional food and neutraceutical source with health promoting benefits. In addition, long shelf life of pomegranate invites huge demand in domestic as well as international market. It is becoming export oriented crop and its area and production are increasing at a faster rate since last two decades. India is considered as one of the largest producers of pomegranate in world having annual production of 1789.31 thousand metric tonnes per annum with 180.64 thousand ha cultivable area and productivity of 9.91 tonnes per ha. Karnataka, in its 23.23 thousand ha area produces 261.82 thousand metric tonnes with productivity of 11.27 metric tonnes per ha. However, pomegranate cultivation with these prospects is largely limited by biotic stresses. Bacterial blight caused by *Xanthomonas axonopodis* pv. *punicae* and lately, wilt caused by *Ceratocytis fimbriata* are reported as major scourge of pomegranate (Anon, 2008).

Integrated disease management schedule is need of the hour for bacterial blight management of pomegranate as there are no resistant sources for the pathogen, whilst; blight management with any single chemical is also difficult. Similarly, for wilt management, curative protocols are very scarce and it is very difficult. In this background, developing the new management schedule for the diseases like bacterial blight and wilt calls for identifying the better IDM components for their use in the different sequences and time points. Pomegranate is an important commercial crop being touted as cash crop and super crop (Sharma and Jadhav, 2011)

Bacterial blight management by integrating the bioagents in IDM

Soil amendment with *Pseudomonas* sp. and *Trichoderma* at the time of pruning and subsequent sprays with formulation containing *Pseudomonas* sp. in different succession alternated with the combinations of Streptocycline (0.5 g/l) + 2-bromo-2-nitropropane-1-3 diol (0.5 g/l) along with micronutrient sprays (Ca, B, Mg, Zn and K) were found recording the least disease whereas chemical treatments with combination of streptocycline 0.5 g and copper oxy chloride had least efficacy over treatment-2 (Benagi and Ravikumar, 2009)

Table 1. Effect of bioagents and other IDM components on disease severity of bacterial blight in pomegranate (Manjunath, 2014)

Treatments	Disease Severity in Different Days					
	60		90		150	
	PDI	Severity	PDI	Severity	PDI	Severity
T1	27.06 (31.34)	11.07 (19.42)	26.42 (30.93)	7.90 (16.30)	24.05 (29.37)	10.10 (18.53)
T2	21.106 (27.34)	6.74 (15.04)	19.10 (25.92)	7.61 (16.02)	14.64 (22.49)	2.74 (9.49)
T3	22.15 (28.07)	8.91 (17.36)	22.19 (28.10)	10.62 (19.01)	18.22 (25.26)	6.53 (14.80)
S.Em±	0.91	2.3	1.2	3.27	0.85	213
C.D (0.05)	1.98	0.99	3.03	1.418	1.66	1.67
CV	12.44	16.52	11.15	10.618	12.10	16.23

Treatments

T1	COC 2 g + Streptocycline 0.5 g as standard check
T2	Soil amendment with consortia of *Pseudomonas* sp., *Trichoderma* sp., and *Aspergillus* sp. with 2 sprays of bioformulations containing *Pseudomonas* sp., at 8 days interval after copper hydroxide (3 g/l) + Streptocycline (0.5 g/l) + 2-bromo-2-nitropropane-1-3 diol (0.5 g/l) (0.5 g/l) along with micronutrient sprays (Ca, B, Mg, Zn and K)
T3	Existing IDM module without Bioagents

Nature of resistance by bacterial antagonists

This experiment was repeated for all the isolates of *Psuedomonas* sp., which were effective against bacterial blight. Challenge inoculation was made in different days from day-1 to day-6 after *Pseudomonas* foliar application keeping each day challenge inoculation as a separate treatment and the mean data is represented. A spatio-temporal pattern experimentation through temporal separation of pathogen and biological treatment revealed that resistance development of 71% was recorded when challenge inoculation was made 3 days after biological treatment application.

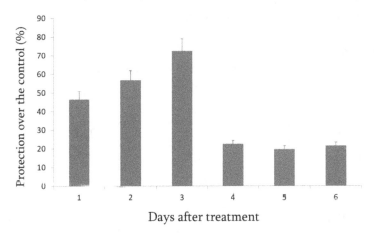

Fig. 1. Spatio-temporal time course studies for understanding the nature of resistance by spatially and temporally separating the bioagents application and pathogen

Table 2. Assessment of agronomic requirements for bioagent's better survival/ tolerance limits in soil after application

Parameters tested	Methods followed	Bioagents tested			
		Trichoderma sp.,	*Pseudomonas* sp.,UHS 1	*Pseudomonas* sp.UHS 11	*Bacillus subtilis*
pH	Potentiometric	4-9	5-8	5-8	6-8
Electrical Conductivity	Conductimetry	3	3	2.5	2.8
Organic carbon (dsm^{-1})	Chromic acid oxidation	0.3-0.4	0.2-0.3	0.3-0.4	0.2-0.3
Minimum of Moisture (%)	Gravimetric method	15-20	14-18	17-20	16-18
ESP%		20-25	8-10	10-13	10-12

Data obtained in the study clearly indicates that *Trichoderma* sp., can survive in the broad range of parameters compared to bacterial antagonists in the soil.

Table 3: Effect of developed consortia along with differential nutrient dosages and organics on fruit yield and its parameters

Treatments	No. of fruits per plant	Fruit weight (kg)	Fruit diameter (mm)	Fruit yield per plant (kg/ plant)	Fruit yield (ton/ hectare
T1- RDF	77[b]	222.5	68.9[b]	17.2[b]	15.14
T2- RDF + consortia	80[b]	283.2	77.3[a]	22.7[a]	19.98
T3-25% RDN org	82[b]	249.1	71.6[b]	20.5[ab]	18.04
T4- 25% RDN org+ consortia	86[ab]	291.2	77.2[a]	24.9[a]	21.91
T5- 50% RDN org	90[ab]	237.5	72.2[ab]	21.2[ab]	18.66
T6- 50% RDN org + consortia	99[a]	263.2	74.1[ab]	25.8[a]	22.70
CD	13.10	NS	5.23	5.3	
Sem±	4.3		1.7	1.7	

Application of developed consortia along with organics, a total of 50% recommended dose of nutrients can be minimized with maximum yield (22.70 ton/ha) obtained. However, the minimum 2-years required for the optimization of this strength in the soil (Greeshma, 2016)

Treatments protocol for reduced mortality in affected orchards and improvement of tolerance against wilt (Manjunath G, 2015)

1nd step: Application of chemical compounds with drenching

Carbendazim 50 WP (3 g)
Fosetyl Al 80 WP (1.25)
                     15 days after
Trifloxystrobin and Tebuconozole (1.25 g)
Choropyrifos -50EC (2 ml)
Boric acid (10 g/plant)
Carbofuran 3G 50 g/tree

2nd step: Application of Humic acid 3 g/l, followed by phosphoric acid 2 ml/l through fertigation or drenching

3rd step: Application of micronutrients ; Potassium silicate, Zinc, Iron, Mg, Boran

4th step   Application 100 g of *T. harzianum* (2x10^8), *P. fluorescens* (10x10^8), *B. bassiana* (2x10^8) and *P. lilacinus* and *A. niger* (25 g/tree) along with 8 kg of well decomposed FYM for 3-feet radius followed by Boric acid 10g/plant

5th step   Prophylactic application of *T. harzianum* 200 g (2x10^8), *P. fluorescens* (10x10^8), *B. bassiana* (2x10^8) and *P. lilacinus* and *A. niger* (100 g/tree) along with 8 kg of well decomposed FYM for 3-feet radius followed by Boric acid

The treatment protocol is being worked out for checking the mortality of plants in wilt affected orchards and validations process is under progress. However, the prophylactic application of *T. harzianum* 200 g (2x10^8), *P. fluorescens* (10x 10^8), *B. bassiana* (2x10^8), *P. lilacinus* and *A. niger* (100 g/tree) along with 8 kg of well decomposed FYM for 3-feet radius followed by Boric acid was found beneficial for improving the tolerance against wilt in addition to good orchard irrigation practices. The given protocol adoption from beginning of planting, maintains the orchard free from wilt incidence.

Summary and Conclusions

Bacterial blight and wilt caused by *Xanthomonas axanopodis* pv. *punicae* and *Ceratocystis fimbriata* are major threats for pomegranate production. Bacterial blight management schedule needed to be developed integrating curative molecules and resistance inducing microbes and SAR agents. The candidate bioagents and consortia of PGPR (plant growth promoting rhizobacteria) and PGPF (plant growth promoting fungi) were found contributing towards enhanced soil health and induction of resistance against blight. Application of bioagents, *Pseudomonas putida* and *Bacillus subtilis* as talc powder and liquid formulations were found beneficial for the management of pomegranate bacterial blight. Prophylactic treatment of bioagents consortia including *Trichoderma harzianum*, *Paecilomyces lilacinus*, *Pseudomonas fluorescens*, *Aspergillus niger* and *Beauveria bassiana* was found to improve the tolerance level of the plant against wilt complex.

References

- Anonymous – NRCP, 2008. Annual report, 2008–2009. *National Research Centre on Pomegranate*, Solapur, pp 36-47.

- Benagi, V. I., and Ravi Kumar, M.R., 2009. Present status of pomegranate bacterial blight and its management. In: Souvenir & abstracts, 2nd international symposium on pomegranate and minor including Mediterranean fruits, org. *ISHS, Belgium*, June 23–27, 2009 at UAS, Dharwad. Karnataka, India, pp 53–58.

- Greeshma, 2016. Effect of bioinoculants and organic supplementation on nutrient content and productivity of pomegranate. Thesis submitted to UHS, Bagalkot. Pp.45.

- Manjunath G, 2014. Technical report of regional horticultural meeting of UHS, Bagalkot.PP.19

- Manjunath G, 2015. Annual report of NRCP, Solapur.pp92

• Sharma, J. and Jadhav, V.T. 2011. *Network Project on Mitigating the Bacterial Blight Disease of Pomegranate in Maharashtra, Karnataka and Andhra Pradesh: Abridged Progress Report*, 2008-09 to 2010-11, National Research Centre on Pomegranate, Solapur, Maharashtra, India, 107

9

Potential of Entomopathogenic Nematodes

S.S. Hussaini[1] and Rajeshwari R.[2]
[1]ICAR- National Bureau of Agriculture Insect Resources, Bangalore
[2]University of Horticultural Sciences, Bagalkot

Introduction

Nematodes parasitizing insects have aroused considerable interest and attention in the light of pressing concerns for the eco-friendly approaches in pest management. As agricultural sustainability is under increasing stress, any promising technique/ methodology is welcome as a saviour of agricultural production, environmental health and natural resources. Entomopathogenic nematodes (EPN) are one such area of eco-friendly methodologies of pest management that is a silver lining in the otherwise pesticide-centric strategies of plant protection by and large.

The history of nematode-parasitism of insects is traceable to seventeenth century AD nothing, significant until 1930s, the report of parasitisation of the grubs of Japanese beetle, *Popillio japonica*. Although, then, Glaser was not aware of the symbiotic bacterium that was associated with the nematodes, yet, he could fortunately devise methods to culture the nematode *in vitro* which were incidentally suitable for the bacterium to multiply. Enough nematodes were produced for field trials to control the beetle. However, much of the lead work done during Glaser's time remained unexplored further until about three decades. With the advent of the era of IPM in the early 1960s, there was renewed interest in EPN as biological control organisms and investigations became important again (Hussaini, S.S., 2002).

Many EPN are known to date but *Steinernema* and *Heterorhabditis* are taxonomically, biologically and commercially most important as they are being seen as the biopesticides of the future. Already, the EPN based biopesticides occupy the major share in the biopesticide world, being second only to *Bacillus thuringiensis* (Bt).

Other EPN such as *Romanomermis culicivorax* have also shown potential in the biocontrol of harmful insects, as mosquitoes, but due to high amounts of labour costs involved in the artificial culture of them (as they survive only on *in vivo* cultures of mosquito larvae), has not been very successful. Use of EPN against the stored grain products was also advocated.

Taxonomic status

Both the economically important genus of EPN viz., *Steinernema* and *Heterorhabditis* belong to the family Steinernematidae and Heterorhabditidae, respectively, in the Order Rhabditida (Class Nematoda). Steinernematidae also contains one monospecific genus, *Neosteinernema longicurvicauda* parasitizing termites.

Biology

a. **Symbiotic Bacteria:** Rhabditida mainly comprises bacteriophagous nematodes and some apparently evolved as insect pathogens. In both families, the infective juveniles (Ijs) are mutualistically associated with two bacterial genera, *Xenorhabdus* and *Photorhabdus*, respectively. *Xenorhabdus* located in specialized intestinal vesicles of juveniles, while *Photorhabdus* in the anterior part of the gut in *Heterorhabditis* juveniles.

Important species of bacteria – Nematode association

Bacterium	Nematode
Xenorhabdus beddingii	*Steinernema longicaudatum*
X. bovienii	*S. feltiae, S. intermedium, S. kraussei, S.affine*
X. nematophilus	*S. carpocapsae*
X. poinarii	*S. glaseri, S. cubanum*
X. japonica	*S. kushidai*
X. indica	*S. thermophylum*

Photorhabdus luminescens sub spp. *luminescens/ laumondii*	*Heterorhabditis bacteriophora*
P. luminescens sub spp. *akhurstii*	*H. indica*
P. temperate	*H. megidis, H. zealandica,* NC group of *H. bacteriophora*

32 species and 76 strains of *Xenorhabdus*, isolated from at least 27 species of *Steinernema* nematodes.

Xenorhabdus species

1. *Xenorhabdus beddingii* (Akhurst, 1986) Akhurst and Boemare 1993 2. *X. bovienii* (Akhurst 1983) Akhurst and Boemare 1993 3. *X. budapestensis* Lengyel *et al.* 2005

2. *Xenorhabdus cabanillasii* Tailliez *et al.* 2006 5.*Xenorhabdus doucetiae* Tailliez *et al.* 2006

3. *X.ehlersii* Lengyel *et al.* 2005 7. *X.griffiniae* Tailliez et al. 2006 8. *X.hominickii* Tailliez *et al.* 2006 9. *Xenorhabdus innexi* Lengyel *et al.* 2005

4. *X. japonica* Nishimura *et al.* 1995 ; 11.*X. koppenhoeferi* Tailliez *et al.* 2006

5. *X. kozodoii* Tailliez *et al.*2006 ;13.*X. luminescens* Thomas and Poinar 1979

6. *X. mauleonii* Tailliez *et al.*2006 ;15.*X. miraniensis* Tailliez *et al.* 2006

7. *X. nematophila* (Poinar and Thomas 1965) Thomas and Poinar 1979 AL

8. *X. nematophila* subsp. *beddingii* Akhurst 1986 ;19.*X. nematophila* subsp. *beddingii* Akhurst 1986 20. *X.nematophila* subsp. bovienii Akhurst 1983

9. *X.poinarii* (Akhurst 1983) Akhurst and Boemare 1993

10. *X.nematophila* subsp. poinarii Akhurst 1983 ;23."*X.nematophilus* subsp. *nematophilus*" (Poinar and Thomas 1965) Thomas and Poinar 1979

11. *X.poinarii* (Akhurst 1983) Akhurst and Boemare 1993

12. *X.romanii* Tailliez *et al.* 2006 ;26.*X.stockiae* Tailliez *et al.* 2006 27.*X.szentirmaii* Lengyel *et al.* 2005

Photorhabdus species

1. *P. asymbiotica* subsp. *asymbiotica* Fischer-Le Saux *et al.* 1999

2. *P. asymbiotica* subsp. *australis* Akhurst *et al.* 2004 VP

3. *P. luminescens* (Thomas and Poinar 1979) Boemare *et al.* 1993 emend. Fischer-Le Saux et al. 1999 VP Syn: *X. luminescens* Thomas and Poinar 1979

4. *P. luminescens* subsp. *akhurstii* Fischer-Le Saux *et al.* 1999

5. *P. luminescens* subsp. *kayaii* Hazir *et al.* 2004

6. *P. luminescens* subsp. *laumondii* Fischer-Le Saux *et al.* 1999

7. *P. luminescens* subsp. *luminescens* (Thomas and Poinar 1979) Boemare *et al.X. luminescens* Thomas and Poinar 1979

8. *P. luminescens* subsp. *thracensis* Hazir *et al.* 2004 *P. temperata* Fischer-Le Saux *et al.* 1999

Association between bacteria and nematodes is mutualism. Nematodes need bacteria for killing the insect host, producing antibiotics that suppresses the growth of other microorganism. It metabolizes the host body contents so that, the nematodes can feed on them and eventually multiply. The bacteria need nematodes for protection from the external environment as bacteria do not possess any resistant stages, facilitating penetration into insect host body.

Pathogenicity and Life-Cycle

Third stage juveniles are the infective stage (Ij3), occur as free-living forms and enter the insect host through any natural opening such as mouth, anus, spiracles or in addition by direct penetration through the inter-segmental membrane. *Heterorhabditis* is able to cut through the cuticle at inter-segmental region to enter inside the host with the help of a dorsal tooth, while the *Steinernematids* can enter the host body only through natural openings. If the juvenile enters by mouth or anus, it penetrates the gut wall to reach the haemocoel, otherwise it reaches the tracheal wall after entering through spiracles. On reaching the haemocoel of the host, the Ijs release bacteria, which multiply rapidly on the haemolymph, resulting in death of the insect within 24-72 h. The symbiotic bacteria are thus responsible for mortality of most insect hosts by septicaemia. The insect dies not only due to the bacterial feeding inside the haemolymph, but also by the toxin produced by nematode. Subsequently, Ijs start feeding on bacteria and their metabolic by-products and moult to the 4th stage and develop into males and females.

The *Steinernematids* develop into amphimictic generation and the *Heterorhabditids* form hermaphroditic females. Females lay eggs that hatch as 1st juveniles and moult successively to 2nd, 3rd and 4th stage juveniles and then to males and females of the second generation. The adults mate and the eggs produce females which hatch 1st stage juveniles that moult to the 2nd stage. The late 2nd stage juveniles cease feeding, reserve a pellet of bacteria in the gut and moult to the 3rd stage (Ijs), which leave the cadaver in search of new hosts. The cycle from entry of *Steinernema* Ijs into a host till emergence of Ijs from a host takes 7-10 days at 25°C in *G. mellonella*. However, life cycle is highly temperature dependent. Life cycle of *S. siamkayai* in *H. armigera* took 11 days to complete development at 30°C (Shapiro et al, 2001).

Survival of Juveniles

Optimum temperature requirements for survival, infection and development vary with the nematode spp. and their original locality. *Steinernematids* remain active at lower temperatures (4-14°C) than *Heterorhabditids* (10-16°C), which may reflect their temperate origins. The duration of nematode survival in the soil in the absence of a host depends upon factors such as, temperature, humidity, natural enemies and soil type. Generally, survival is better in sandy soil /sandy-loam soil at low moisture and with temperature, 15-25°C than in clayey soils and at lower or higher temperatures. *Heterorhabditids* do not survive as efficiently as *Steinernematids*. Ijs do not feed and survive for months on stored carbohydrate reserves in an anhydrobiotic state. *Heterorhabditids* are resistant to rapid desiccation. Ijs of *S. tami* and *S. carpocapsae* survived up to 40 days in distilled water without mortality or loss in infectivity while *S. abbasi* recorded 2% mortality. Effect of ageing of Ijs on pathogenicity was not altered in stored and not stored *Steinernema* spp. whereas fresh Ijs were non-infective compared to stored Ijs of *Heterorhabditis* spp.

Factors Affecting Pathogenicity

Biotic and abiotic factors affect the biology and pathogenicity of nematodes. Six different temperatures (10-35°C at 5°C intervals) were studied to determine the insecticidal activity of *S. kushidai* against beetle, *Anomala cuprea*. Mortality was 95, 100 and 85% at 20, 25 and 30°C respectively, and significantly lower at 10, 15 and 35°C. Nematode distribution, persistence and virulence in the soil was affected by soil type, texture and structure and also with varying water potential in soil. Effect of metal ions (Al, Cd, Co, Cr, Cu , Fe, Li, Mg, Mn, Mo, Ni, Pb, Se, V and Zn) on the mortality and pathogenicity of *H. bacteriophora* and *S. carpocapsae* showed with the exception of Pb (II), other ions did not cause any mortality to the nematodes. However, Ni and Pb, reduced their pathogenicity.

Dispersal of juveniles

Juveniles are capable of dispersal vertically and horizontally, actively and passively dispersed by rain, wind, soil, humans or even insects under mere entomophilic association. Active dispersal may be upto a few centimetres and passive dispersal by insects may be upto several kilometres. They are quite capable of long-distance dispersal and local migration. Transmission strategies include both highly active cruisers and ambushers. Studies on dispersal of EPN by earthworms, *Lumbricus terrestris* showed that upward dispersal increased in presence of earthworms and downward dispersal of *S. carpocapsae* and *S. feltiae* was not observed.

Competition with other organisms and natural enemies

The dynamics of interactions of pests and natural enemies plays an important role in the population ecology of all organisms. Populations of EPN are affected negatively through biotic stresses such as bacteria, fungi, predatory nematodes, tardigrades, mites, etc, where in mites are most effective in reducing their populations. Hence, the survival is better in sterilized than in non-sterilized soil.

Host Range

EPN are reported to infect over 200 insect species. The insects killed or parasitized include, armyworms, carpenter worms, cat fleas, crown borers, cutworms, filth flies, flea beetles, German cockroaches, leaf miners, mole crickets, phorid flies, plume moths, root weevils, sciarid flies, stem borers, webworms, white grubs etc.

Differential susceptibility of life-stages of target pests

With the understanding that certain developmental stages of pests are more susceptible to EPNs, the effects of *Heterorhabditis* sp. (HNI-Cenicafe) and *S. feltiae* (Sf-Villapinzon) on the mortality of different stages (L_1, L_2, L_3 young, L_3 mature and pre-pupae), of two white grub species, *Phyllophaga menetriesi* and *Anomala inconstans* were studied. The strains caused different mortality rates on *A. inconstans*, with higher mortality for *Heterorhabditis* than *S. feltiae* for different instars, although L_2 was more susceptible to the first strain. The greatest mortality in *P. menetriesi* occurred at 20 days after inoculation with HNI, where L_2 was also the most susceptible stage (81.1%). Hence, susceptibility of white grubs to EPNs depends on the host species and EPN strain used. *S. carpocapsae* (Mexican) was reported as the most pathogenic species, followed by *H. bacteriophora*, *S. feltiae*, *S. anomali* and *S. glaseri* against *Diaphania hyalinata*.

Differential virulence and development of *S. carpocapsae* (Mexican), on different life-stages of the western corn rootworm, *Diabrotica virgifera* was demonstrated. The 3rd instar rootworm larvae were 5 times more susceptible to nematode infection than 2nd instar larvae and 75 times more susceptible than 1st instar larvae and pupae. Rootworm eggs were not susceptible. Nematode development was observed in all susceptible rootworm stages, but a complete life cycle was observed only in 2nd and 3rd instar larvae and pupae. Nematode size was affected by rootworm stage. The least number of Ijs were recovered from 2nd instar rootworm larvae and *S. carpocapsae* was to be applied when 2nd and 3rd instar rootworm larvae were predominant in the field.

Effect on non-target organisms

Biological control through nematodes has often been advocated to be ecologically safe. Only a few studies deal with the action of EPN on non-target animals, although a broad spectrum of species have been tested in the lab. EPN do not affect vertebrates under natural conditions. Mortality can occur to non-target arthropod populations, but will be temporary, spatially restricted and affecting part of a population. In EPN treated, the impact on the non-target fauna was negligible. The relatively short period of persistence of EPN and the necessity of their populations to recycle frequently in hosts make it unlikely that, they could have major effects on non-target organisms. Their selectivity and beneficial traits as biocontrol agents outweigh the small risks of causing unwanted environmental disturbance in non-target populations. Non-susceptibility of non-target hosts such as honey bees is on record. *Apis mellifera*, worker bees and brood when sprayed with four EPN spp. on to the combs resulted in complete resistance (Hussaini and Singh, 1998).

Effect of host plant on biocontrol potential

Assays were conducted to determine the effects of host plant and a plant secondary metabolite, cucurbitacin D on the mortality of *Diabrotica undecimpunctata howardi* from infection by EPN. Rootworms produced on maize, groundnut and two squash varieties, one containing cucurbitacin D and the other lacking, were exposed to different EPN strains. Rootworms fed on maize suffered lower mortality than those reared on groundnuts or either of the squash varieties. Rootworms fed on the squash varieties suffered greater mortality. Multiplication was highest from rootworms fed on maize, lower for groundnut and lowest on squash. Progeny production from rootworms fed on bitter squash was lower than from non-bitter squash for all nematode strains teste. This attributed to the effects of cucurbitacins and other plant primary metabolites on rootworms and EPN.

Practical Advantages of EPN

Attributes of good EPN biocontrol agents

To be successful biocontrol agent, EPN have to exhibit certain ecological attributes. An affordable cost-benefit ratio is essential for the practicable solution. Environmental tolerance and adaptation especially in case of foliar sprays of EPN, is an essential attribute. Good dispersal after their application, passive or active, is important. A combination of ambushers and cruisers is applied together. Seasonal coincidence of all the three i.e., the parasite, the insect and crop life-cycles have to coincide to be successful biocontrol agents. Physiological acceptance by the host i.e., the host should not have immunity against the EPN. Competitively, nematode should multiply in host and environment so that repeated applications are not required. As they qualify most of these traits, they are generally regarded as very good biocontrol agents (Hussaini et al., 2017).

Main strongholds of EPN to be successful biocontrol agent

Like parasitoids and predators, EPN has chemoreceptors. Like pathogens, they are highly virulent killing their host within 24-48 h., cultured easily *in vitro*, has high reproductive potential and broad host range. Are safe to the plant, animals and the environment, can be easily applied using standard spray equipments, has potential to recycle in the environment and are compatible with many chemical pesticides.

Applied aspects: Compatibility with pesticides and other biocontrol agents

EPN have advantages over chemicals as biocontrol agents. They are non-polluting and thus environmentally safe and acceptable, although some countries do not allow the release of non-indigenous species. Infective juveniles can be applied with conventional equipment and they are compatible with most pesticides. They find their hosts either actively or passively and in cryptic habitats and sometimes in soil. They have been proven superior to chemicals in controlling the target insect as per Gaugler, (1981). They are not well suited for foliar application since they are sensitive to desiccation and UV radiation. The effective host range of a given species or strain is usually rather narrow, thus they do not cause indiscriminate mortality. The narrow host range means that one must select the appropriate nematode just as one must select the appropriate chemical insecticide to control the target insect.

Combined use of 48% Chlorpyrifos EC (1000 mg/lit), 70% Imidacloprid (500 mg/lit) and *S. carpocapsae* (4,000 Ijs/ml) against *Rhabdoscelus lineaticollis* on palms and sugarcanes, showed 98% mortality after 7 days which was higher than individual treatments with chemicals and *S. carpocapsae*. Mortality of the weevil adults in the combined treatment was higher. 2% w/v concentration of neem seed kernel extract (NSKE) affected 3rd instar larvae of *Heterorhabditis* spp. causing mortality while *Steinernema* spp. were less susceptible. Treated *Steinernema* spp. killed all *G. mellonella* larvae while treated *Heterorhabditis* spp. were unable to parasitize. *Steinernema glaseri* had higher survival, recovery and maintained infectivity after pesticide exposure than *Heterorhabditis* spp. Interspecific competition of EPN with *B. thuringiensis* (*Bt*) is dependent upon the timing of *Bt* exposure to the host. Lepidopteran host infected with EPN 24 h prior to *Bt* have normal nematode development, while nematode growth was poor or negligible when *Bt* was inoculated 24 h prior to EPN. When both were inoculated simultaneously, a few cadavers exhibited dual infection, with *Bt* occupying the anterior part, and nematode, the posterior part of insect. Nematode growth was however poor in dual infection.

Efficacy of *B. thuringiensis tenebrionis* (Novodor-FC) (*Btt*) protected elm (*Ulmus americana*) foliage from damage by elm leaf beetle, *Xanthogaleruca luteola* [*Pyrrhalta luteola*]. Untreated trees lost up to 40% of their total foliage due to leaf beetle feeding in 3 weeks, while *Btt* treated trees suffered 10% defoliation. *Steinernema carpocapsae* incorporated into tree bands containing cellulose mulch, proved effective at killing high proportions of migrating larvae.

Interactions between *H. bacteriophora* and *Metarhizium anisopliae* with respect to mortality, production of Ijs and production of conidia were studied during dual infections of sugar cane borer, *Diatraea saccharalis*. A positive effect was demonstrated for host mortality in dual infections. Results showed that, for faster time to death, a moderately virulent fungal isolate could be combined with the nematode at the expense of Ijs production. *Beauveria bassiana*, when inoculated simultaneously on *G. mellonella* with *S. carpocapsae* or *H. bacteriophora*, were suppressed however, the nematodes had an antagonistic effect when the fungus was applied prior to the nematodes.

The effects of the agitation system, initial temperature of the spray liquid, EPN concentration and additional air injection on the viability of EPNs showed that the hydraulic agitation caused more reduction in viability than the mechanical agitation. A lower temperature of the initial spray liquid yielded higher viable EPN compared to a higher temperature after hydraulic mixing and so did air injection, while EPN concentration did not significantly influenced viability.

Post application survival

Five distinct phases have been identified and each phase being associated with a specific set of mortality factors. Pre-application factors are with production, storage and transport conditions governing survival rate and quality of nematodes at application time. The next phase of tank mixing and application with a sprayer etc. usually does not cause mortality as Ijs are quite tolerant of shear forces. The most critical for survival are the first few minutes and hours after application leading to high mortality (40-80%). UV radiation and dehydration are the most important mortality factors. The remaining nematodes settle in the soil and their numbers gradually decrease (5-10% per day). Predation, infection by antagonists, depletion of energy and desiccation are probably the main factors during this period. In most cases, after 2-6 weeks less than 1% of the applied population is still alive. Through recycling in host insects, nematodes may persist for years at these levels. Hence, the pattern is a rapid decline in the first few days followed by a moderate decline over the next 2-6 weeks and then a long period of recycling at a low level (Hussaini et. al., 2014).

Persistence of EPN in sandy loam soil with variable moisture levels showed that

H. bacteriophora and *H. zealandica* persistance was short at -10 kPa, improved slightly at -100 kPa, significantly at -1000 kPa and was highest at -3000 kPa. *Steinernema scarabaei* and *S. glaseri* persisted very well at -10 kPa.

Strategies for enhancing survival of EPN

Since the survival of EPN under the stressed conditions of temperature and humidity are the major challenges, there have been numerous surveys to isolate strains which are more tolerant to such stresses. Besides, some novel strategies such as adjuvant chemicals have been tried to make the EPN more tolerant to these stresses. Glycerine is most effective adjuvant for increasing the mortality of *L. trifolii* by *S. carpocapsae* strain sprayed on chrysanthemum under conditions of high humidity in plastic cages. Free moisture on the leaf surface and increased RH enhanced the survival and activity of EPN. Induction of anhydrobiosis through slow desiccation of three strains of *S. feltiae* was demonstrated and were able to induce a quiescent anhydrobiosis state in all strains, which enabled them to survive at lower RH (75-85%) in Israel. Infective ability of the EPN against insects in sea water (salinity 38 ppt) with elevated temperature in aqueous suspension was studied and found that Ijs of *H. indica* tolerated more salinity at high temperature (40 °C) in seawater than in distilled water but infectivity was 15% (Hussaini et al., 2009).

Culture and mass production

EPN can be reared *in vivo* in insect hosts or mass produced *in vitro* on solid or liquid medium. For solid medium culture, substrate such as beef or pork kidney, liver or chicken offal was used. Substrate was made into a paste and coated onto a porous substrate such as sponge, medium was sterilized, inoculated with the bacterium and nematodes were added after 24 h. Infective juveniles were harvested after 15 days. This method is labour intensive but well suited for cottage industry where labour is plentiful. Production in liquid medium was done in small containers or in fermentation tanks. Greater numbers of juveniles were produced per unit area in fermentation tanks, which makes this method especially suited for large-scale commercial production.

Economics of EPN use

Estimated that it costs 10-60% more to control insects with nematode-based products than with chemical insecticides. As technological improvements in production, formulation, packaging and shelf life of nematode products occur, some clients opt for a biological control method at a higher cost due to safety to environment. Higher short-term cost may be lower in the long run when continued control by the recycling nematode is obtained. Thus, by the judicious use of nematodes and chemicals, it may be possible to reduce the cost of control and protect the environment at the same time. An excellent example is the control of black vine weevil, *Otiorhynchus sulcatus* in cranberries. Chemical insecticides on cranberry was restricted or provided no adequate control. The application of *H. bacteriophora* NC applied provided more than 70% control up to one year (Hussaini et al., 2005).

Control of *Plutella xylostella* (DBM) using polymer formulated *S. carpocapsae* and *Bt* in cabbage fields (Java and Indonesia) was demonstrated. A single use of 0.5 million Ijs of *S. carpocapsae* m-2 applied with a surfactant polymer formulation containing 0.3% xanthan and 0.3% Rimulgan achieved a significant reduction of the insects per plant with >50% control after 7 days. Weekly applications of *B.t* (TurexReg.) or alternating applications of TurexReg. and the nematodes achieved >80% control.

Guidelines for field application

1. Soil temperature at application time (preferably even after application), should be 18-30 °C.

2. Nematodes should be applied to moist soils. Pre or post-application of irrigation and continued moderate moisture is necessary.

3. Application should be done preferably in the morning or evening to avoid exposure of EPN to UV radiation and high temperature.

4. High spray volumes to be applied to reach optimum depth.

5. Dosage rate of 2.5-7.5 billion/ha is consistent in the fields for pest control. About 50 EPN/flying insect are sufficient. Repetition of application is required for insects like white grubs.

6. Four week duration of trial is needed normally. For Lepidopteran insects, 3 to 7 days duration is ideal. Flood jet nozzle is fairly effective for spraying.

Application Methods

Spraying: Like chemical insecticides, spraying nematodes directly on to the soil surface/ foliage is the most common method.

Trickle irrigation: Button type of trickle irrigation outlet is used to deliver the nematodes to the area on one side of the plant.

Capsule: Prepared from wheat bran (5% w/w) with calcium alginate containing 1,000-2,000 nematode/capsule is applied. Capsules are buried in soil or under dairy compost. 70-80 capsules / plant are used.

Liquid Baits: Desiccated nematodes (*S. feltiae*) are mixed with 56% sucrose solution and used. **Pellet Baits**: Wheat bait pellets from wheat bran and wheat flour, locust bean gum (18mg/ml water) and corn oil are used.

Nylon pack cloth bands: Nematodes are applied to nylon pack cloth bands lined with fleece or terrycloth that is wrapped around tree trunks to control gypsy moth larvae, *Lymontria dispar*.

Cardboard band: *Steinernema carpocapsae* incorporated into cardboard band placed around the trunk of apple trees as an artificial bark substrate, infected 23-73% of codling moth pre-pupae moved under the barks.

Punch and Syringe method: In forest trees, inoculation hole is made by hammer and delivery of about 1 ml of nematodes containing medium (aerated 12% gelatin with 4,000 nemas/ml) is done by syringe.

Dosage and frequency of application

Steinernema carpocapsae was inoculated to squash (*Cucurbita pepo*) twice per week, @ 3 billion to control pickleworm, *Diaphania nitidalis*, on cucurbits in Florida. Fruit damage was 0-9% in treated and 33%-60% in untreated plants. Blossom damage was also reduced. Even 1 bill/per ac were as effective as permethrin, although neither treatment was completely effective when pickleworms were abundant. More frequent applications may be necessary to achieve control with a reduced rate of nematodes. Different researchers have put the ideal dosage rate at 1-7.5 bill/ha.

Commercial products

Most of the nematode based products currently available are formulations of various strains of *S. carpocapsae* such as ORTHO BioSafe, BioVector and Exhibit in the USA, Sanoplant (Switzerland), Boden Niitzlinge (Germany) and Helix (Canada) (Table-III). Other species of *Steinernema* commercially available are *S. feltiae* as Magnet in US and as Nemasys and Stealth in UK, *S. riobrave* as Vector MC and *S. scapterisci* as Proactant Ss in US. *H. bacteriophora* is available as Otinem in US and *H. megidis* as Nemasys in UK.

EPN Research in India and in other warmer climates

Reports on the occurrence of important species of EPN and small success stories since 1960's are available. A steady progress in EPN research has been undertaken over the past one decade. EPN research in India dates back to 1966 when work on DD136 strain of *S. carpocapsae* was considered against the pests of rice, sugarcane and apple. Life cycle of DD136 strain, compatibility with insecticides and fertilizers and lab tests on bioefficiency was studied at CRRI, Cuttack. Later, EPN were reported on *Agrotis ipsilon* and *A. segetum* in irrigated potato fields at Punjab. *Heterorhabditis indica* was reported on sugarcane top borer *Scirpophaga exerptalis* from Coimbatore and *S. masoodi*, *S. seemae* and *S. qazii* was reported from pulses. *S. glaseri*, *S. riobrave*, *S. siyamkaii* has been reported later. Records of EPN have been from Kerala and Tamil Nadu, Uttaranchal, Rajasthan, Assam, Delhi, Gujarat, HP. Uttar Pradesh, Haryana and Meghalaya. However, field data on their use is limited. Although, there are several reports on the efficacy of the EPN against important pests such as *H. armigera*, *Rhynchophorus ferugineus*, *Plutella xylostella*, *Holotrichia serrata*, *A. ipsilon*, *Mylloceros discolor*, *Leucinodes orbonalis*, *Achaea janata*, *Cnaphalocrocis medinalis*, *Galleria mellonella*, *Oryctes rhinoceros*, *Pericallia ricini*, *Rhynchophorus ferrugineus*, *Spodoptera litura*, *Odoiporus*, *Plutella* and *Chilo partellus*. Systematic surveys were initiated by Project Directorate of Biological Control (PDBC) in 1996 throughout the Country which led to documentation of several new isolates and species (Hussaini, S.S., 2014).

In early 1990's, two products namely, 'Soil Commandos' and 'Green Commandos', were marketed but they were not effective as the nematodes carried in the formulation were of European origin and they failed to acclimatize under Indian conditions. It is crucial to identify and select pathogenic native nematode species/strains, which can adapt to the ecology and biology of the insect hosts prevalent in similar agro-climatic zones. Biological control of insect through EPN has not been taken up too enthusiastically in warmer countries especially in Asia. Workers have concentrated on the description of new species and their biocontrol potential has been studied mostly under lab conditions.

Since, most of the work on EPN has been conducted in temperate countries where the climatic conditions are totally different from the tropical or subtropical countries, the nematodes fauna is so divergent that the EPN could hardly be exchanged without any risk of their suitability in complementary situations.

Symbiotic bacteria as a model organism

Photorhabdus is considered to be a more broad-spectrum insect pathogen and is the focus of more research. The complete genome sequence of *P. luminescens* subsp. *laumondii* strain TT01 a symbiont of *H. bacteriophora* isolated on Trinidad and Tobago, has been deciphered which constitutes of 5,688,987 base pairs containing 4,839 protein coding genes, 157 pseudogenes, seven complete sets of rRNA operons and 85 tRNA genes. The protein coding genes are predicted to encode a large number of adhesions, toxins, heamolysins, proteases and lipase and contain a wide array of antibiotic synthesizing genes. These proteins are likely to play a role in the elimination of competitors, host colonization, invasion of insect host and bioconversion of the cadaver. These aspects make *Xenorhabdus* and *Photorhabdus* a promising model for the study of symbiosis and host-pathogen interactions.

International status of Biosafety regulations

The commercial development of EPN gained momentum in the 1980s in USA and Canada since these were exempted from registration under the EPA. In addition to the indigenous EPN, exotic species like *S. scapterisci* from S. America and *S. feltiae* from Europe were permitted entry. Guidelines for introduction of EPN were suggested such as need to export, foreign exploration, taxonomy, shipment, quarantine, permits, host range tests, documentation etc. but, these did not become official. Biosafety concerns were also raised in USA.

In Europe, the OECD concluded that the EPN are safe and should not be subject to registration. However, introduction of antagonistic EPN should be regulated. In U.K., the indigenous unmodified EPN and their symbiotic bacteria are not subject to registration but, introduction of exotic ones is strictly under Wildlife Conservation Act, 1982. The European Commission directive requires limited registration procedure which is only a paper exercise in countries like Sweden, Norway and Ireland, but in Germany and France, a phytosanitary certificate is insisted.

In Japan, Health and Agriculture departments, insist for EPN registration. Better efficacy of a new product must be demonstrated for 2 years in field trials and toxicity studies of nematodes and symbiotic bacteria should be carried out by an independent lab. However, Japan considered nematodes as benign and speeded up registration at a reduced cost. Australia considers EPN as macro-parasites exempt from registration but introduction of exotic EPN is regulated through the Quarantine authority for import permits. New Zealand regards them as any other pesticides that must be registered before trials and sale. Hence, wide variance is noticed but largely considered safe and restrictions against native ones are none or nominal.

Future role: The future of nematode-based products for insect control is very promising. The technology used currently for mass production, formulation, storage and transport has developed much over the time. However, there is still considerable level of information on the biology and potential efficacy, of especially indigenous strains, which need to be worked out. Even though a number of species have been described, only some are commercialized such as, *S. carpocapsae*, *S. feltiae*, *S. riobrave*, *S. scapterisci*, *H. bacteriophora* and *H. megidis*. The role of developed countries like Australia and Europe and North America is, they could make significant contributions for the development of technical know-how. However, more focused efforts are to be made for the exploration of indigenous populations, which can cater to the local needs more effectively. Chemicals such as brighteners, which protect nematodes from harmful ultraviolet radiation and anti-desiccants could be devised in future for their utilization. Currently, the share of the pest control market captured by EPN is very low, that is less than 2% but is likely to increase in future. As the user fully realizes that nematodes are biological organisms and must be handled as such to provide effective control, greater acceptance of nematode based products will occur. Should the availability of chemical pesticides decline sharply, the use of nematode products could expand and fill the void? For control of certain insects, nematodes may replace chemical pesticides and in other cases, they will be used in conjunction with them. Though it is largely unlikely that EPN will ever dominate most pest control scenarios, they definitely have a place of their own and their scope is definitely very bright. In countries like India, their success is bound to be there with the incorporation of EPN methodologies in relevant IPM programmes and with the farmers' active participation, especially in the mass production of EPN.

References

- Gaugler, R., 1981. The biological control potential of neoapectanid nematodes. Journal of Nematology, 13 (3): 241-9

- Grewal, P.S. and Georgis, R., 1988. Entomopathogenic nematodes. In: Methods in Biotechnology. Hall, F.R, and Menn, J.J., (Eds.), Vol. 5, Biopesticides: use and delivery, Totowa (NJ): Humana Press Inc. pp. 271–297.

- Hussaini, S.S., 2017. Entomopathogenic Nematodes: Ecology, Diversity and Geographical distribution, p.88-142 *In* "Biocontrol Agents - Entomopathogenic Nematodes and Slug parasitic Nematodes", MMM Abd Elgawad, T.H. Askary, James Coupland (Eds.), CAB International, Oxfordshire, UK.

- Hussaini, S.S., 2014. Potential of Entomopathogenic Nematodes in Integrated Pest Management *In: Integrated Pest Management - Current Concepts and Ecological Perspective.* Dharam P. Abrol (Eds.), Academic Press, NY, pp.193-223.

- Hussaini, S.S., 2009. Potential of Entomopathogenic Nematodes for the suppression of Insect Pests, *In: Biopest Management - Entomopathogenic Nematodes, Microbes and Bioagents* (Eds.), H.C.L., Gupta, A.U., Siddiqui, Aruna Parihar, Agrotech Publishing Academy, Udaipur- 313 001. Rajasthan, pp. 13-38.

- Hussaini, S.S., 2005. Formulations of Entomopathogenic nematodes, *In: Microbial biopesticide formulations and application*, R.J. Rabindra, S.S. Hussaini and B. Ramanujam (Eds.), Project Directorate of Biological Control, Bangalore, pp.159-167.

- Hussaini, S.S. and Singh, S.P., 1998. Entomophilic nematodes for insect control, *In: Biological Suppression of Plant Diseases, Phytoparasitic Nematodes and Weeds*, S.P. Singh and S.S. Hussaini (Eds.), Project Directorate of Biological Control, Bangalore 560 024, Karnataka, pp. 238-267.

- Kulkarni, N., Paunikar, S., Hussaini, S.S. and Joshi K.C., 2008. Entomopathogenic Nematodes in insect pest management of forestry and plantations crops: An appraisal. *Indian Journal of Tropical Biodiversity*, **16(2)**: 155-166.

- Shapiro-Ilan, D. I., Lewis, E. E., Behle, R. W. and McGuire, M. R., 2001. Formulation of entomopathogenic nematode-infected-cadavers. Journal of Invertebrate Pathology, **78**: 17–23.

Table 1. List of species of genera, Steinernema and Heterorhabditis

Genus : Steinernema Travasos, 1927 (98)

species	Year	origin	species	year	origin	species	year	origin
1.S. kraussei	1927	Czech	34..S.thanbi	2001	Vietnam	67.S.texanum	2007	USA
2.S. glaseri	1929	Brazil	35..S.sangi	2001	Vietnam	68.S. costaricense	2007	Costa Rica
3.S.feltiae	1934	NZ	36.S. diaprepesi	2002	USA	69.S.columbiense	2008	Colombia
4.S. carpocapsae	1982	USA	37.S. asiaticum	2002	Pakistan	70.S.ichmusae	2008	Italy
5.S.affine	1982	Europe	38.S. websteri	2003	China	71.S.cholanense	2008	China
6.S. rarum	1988	Argentina	39.S. weiseri	2003	Czech	72.S.australe	2009	Chile
7.S.kushidai	1988	Japan	40.S. scarabaei	2003	USA	73.S.qazi	2009	India
8.S.arenarium	1988	USSR	41.S.anatoliense	2003	Turkey	74.S.xueshanense	2009	China
S.intermedium	1988	USA	42.S.ethiopense	2012	Ethiopia	75.S.litorale	2004	Japan
9. S. ritteri	1990	Argentina	43..S.loci	2001	Vietnam	76.S.boemarii	2009	France
10.S.scapterisci	1990	USA	44.S.hermaphroditum,	2004	Indonesia	77.S.unicornum	2010	Chile
11.S.longicaudatum	1992	China	45..S. yirgalomense	2004	Ethiopia	78.S.braziliense	2010	Brazil
12.S.neocurtillae	1992	USA	46.S.apuliae	2004	Italy	79.S.arasbanense	2011	Iran
13.S.peurtoricens	1994	Peurto rico	47.S.guangdongense	2004	China	80.S.pui	2011	China
14.S. riobrave	1994	USA	48.S. silvaticum	2005	Germany	81.S.vulcanicum	2011	Italy

Species	Year	Country	Species	Year	Country	Species	Year	Country
15.S.bicornutum	1995	Yugoslavia	49.S. beddingi	2005	China	82.S.everestense	2011	Nepal
16.S.cubana	1994	Cuba	50.S. akbursti,	2005	China	83.S.lumjungense	2011	Nepal
17.S.oregonense	1996	USA	51.S. masoodi	2005	India	84.S.phyllophagae	2011	USA
18.S.karri	1997	Kenya	52.S. seemae	2005	India	85.S.meghalensis	2012	India
19. S. siamkayai	1997	Siam	53.S.robustispiculum	2004	China	86.S.xinbinense	2012	China
20.S.abbasi	1997	Oman	54.S. epokense	2006	Vietnam	87.S.ethiopiense	2012	Ethiopia
21.S.monticolum	1997	Korea	55.S.epokense	2006	Vietnam	88.S.changbalense	2012	China
22.S.ceratophorum	1997	China	56.S. hebeiense	2006	China	89.S.dharani	2012	India
23..S.intermedium	1998	USA	57.S. koisanae	2006	S.Africa	90.S.sachbarai	2014	SA
24. S. jollieti	1999	USA	58.S. sichuanense	2006	China	91.S.citrae	2014	SA
25. S.feltiae		USA	59.S. ashiuense	2006	Japan	92.S.poinari	2014	Czech
27.S.tami	2000	Vietnam	60.S. backanense	2006	China	93.S.tophus	2014	SA
28.S.thermophilum	2000	India	61.S. leizhouense	2006	China	94.S.innivationi	2015	SA
29.S.pakistanensis	2001	Pakistan	62. S.seasonense	2006	Costa Rica	95.S.balochense	2015	Pak
30.S.fabii	2016	SA	63. S. peruvianense	2007	do	96.S.jaffraense	2016	SA
31.S.goweni	2016	Venezuela	64.S.sukhotense	2017	India	97.S.pawamensis	2016	Tanzania
32.S.minutum	2010	Tailand	65.S.nguyeni	2016	SA	98.S.tielingense	2012	China
33.S.beilcehmitel	2016	SA	66.S.khyongi	2018	USA			

Genus :*Heterorhabditis* Poinar, 1976 (26)

1.*H. bacteriophora*	1976	Australia	2.*H.zealandica*	1990	Australia	3.*H. megidis*	**1987**	USA
4.*H. indica*	1992	India	5.*H. argentinensis*	1993	Argentina	6.*H. brevicaudis*	1994	China
7.*H. hawaiiensis*	1994	Hawaii	8.*H. marelata*	1996	USA	9.*H. bopta*		USA
10..*H. taysearae*	1996	Egypt	11.*H. downesi*	2002	Wexford	12.*H. baujardi*	2003	Vietnam
13.*H. mexicana*	2004	Mexico	14.*H.floridensis*	2006	USA	15.*H. amazonensis*	2006	USA
16.*H. hepialis*		USA	17.*H. beliotbidis*	1976	Aukland	18.*H.sanorensis*	2009	Mexico
19.*H. gerrardi*	2009	Australia	20 *H. beicberriana*	2012	China	21.*H.atacamensis*	2011	Chile
22.*H.noenieputensis*	2014	SA	23.*H. safricana*	2008	S.Africa	24. *H eterorbab ditidoides rugaoensis*	2012	China
25.*H.mexicana*	2004	Mexico	26.*H.georgiana*	2008	USA			

10

Recent Advances in Mass Production, Formulation and Field Application of Entomopathogenic Nematodes

Jagadeesh Patil and R. Vijayakumar
ICAR-National Bureau of Agriculture Insect Resources, Bangalore

Introduction

With the rapid development and advancement of synthetic chemistry in the 1930s and by the early 40s, a range of new pesticides had been developed. These potential compounds were used as insecticides to control insect pests. Use of chemical insecticides has been necessary to enhance economic potential in terms of increased production of food, fibre and scaling down of vector-borne diseases in crops. However, indiscriminate use of chemical insecticides can destroy the beneficial natural enemies; induce resistance in pest populations, pesticidal residues in food chain and serious health implications to man and his environment.

Therefore, any attempt to scale down the use of chemical insecticides is welcome considering and the safety to flora and fauna of the environment. Due to environmental and regulatory pressures, use of potential biocontrol agents has safer and alternative tool for pest management. Biological control exploits insects, bacteria, viruses, fungi and nematodes as biological insecticides. Among them, Nematodes are the most numerous multicellular animals on earth. They are microscopic, non-segmented, elongated round worms with lacking appendages. Nematodes can be found in marine, fresh water, and terrestrial environments. They do exist in the environment, as free-living as well as parasitic to plants and animals including human beings.

Among the various parasitic nematodes, some beneficial nematodes do exist in soil environment. These nematodes have a symbiotic association with insect pathogenic bacteria and ability to kill insect pests. These nematodes are generally known as "entomopathogenic nematodes" (EPNs).

Isolation of Entomopathogenic Nematodes

Isolation of EPNs, the random sampling is generally employed when focusing on an extensive area. This sampling strategy has been used for studying the diversity of EPN from a large geographic area or region. A number of factors can be considered depending on the focus of the study including a diverse range of elevations, soil textures and habitats (e.g., cultivated fields, forests, pastures, parks, seashores, riparian areas, etc.).

1. **Soil sample collection**

    1. At least 2-4 m^2 area should be covered for each sampling site

    2. Soil samples should be collected at a depth of at least 15-20cm.

    3. Take at least 5 random samples within this area and at least 3 subsamples should be drawn per sample. Then combine the subsamples as one composite sample.

    4. Place each sample in a plastic bag and label samples with a waterproof marker with the following information from each sampling site: Site information (including area/ site name and GPS coordinate information), date, type of vegetation, soil type, elevation, etc.

    5. Clean collecting tools between samples by thoroughly washing with water.

    6. Samples should not be exposed to sun light (temperature should not be >20°C) during transport to the laboratory.

    7. Take a portion of the soil sample for analysis to obtain information on soil composition, texture, moisture, electrical conductivity, organic matter content etc.

NOTE: Better to collect and/or record the presence of insects or other invertebrates in that location. They may represent potential natural EPN hosts.

Nematode Isolation from Soil Samples: Insect-baiting technique and recovery of EPNs from Infected Cadavers by Modified White Trap

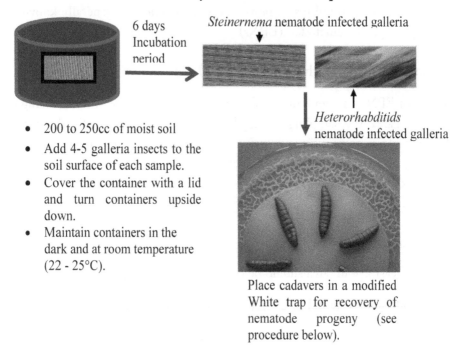

6 days
Incubation
period

Steinernema nematode infected galleria

Heterorhabditids nematode infected galleria

- 200 to 250cc of moist soil
- Add 4-5 galleria insects to the soil surface of each sample.
- Cover the container with a lid and turn containers upside down.
- Maintain containers in the dark and at room temperature (22 - 25°C).

Place cadavers in a modified White trap for recovery of nematode progeny (see procedure below).

Culturing of Entomopathogenic Nematodes

The approach consists of inoculation, harvest, concentration, and (if needed) decontamination. Insect hosts are inoculated on a dish or tray lined with absorbent paper or another substrate conducive to nematode infection such as soil or plaster of Paris. After approximately 2-5 days, infected insects are transferred to the White traps; if infections are allowed to progress too long before transfer, the chance of the cadaver rupturing and harm to reproductive nematode stages is increased (Shapiro-Ilan *et al.*, 2001).White traps consist of a dish or tray on which the cadavers rest surrounded by water, which is contained by a large arena. As IJs emerge they migrate to the surrounding water trap where they are harvested. The scale of the White trap in size and number can be expanded to commercial levels.

Mass production of Entomopathogenic Nematodes

EPNs are easily mass produced by two methods 1) *in vivo* and 2) *in vitro* (solid and liquid culture).

1. *In vivo* culture method

The *in vivo* production of EPNs required an insect host. The wax moth larva, *Galleria mellonella* is the insect of choice for *in vivo* production because it can be mass multiplied on artificial diet, large body mass and easy to handle, whereas an artificial media are used in *in vitro* method. Each approach has its merits and demerits in relation to production cost, technical knowledge, product quality and cost of production and also each method has the potential for improvement.

2. *In vitro* culture method

In vitro culturing of these EPNs required a detailed understanding of the behaviour and biology of the different nematode species. This method is based on introducing nematodes to nutritive medium containing their specific symbiotic bacterium. *In vitro* culturing can be done by two methods. The *in vitro* liquid culture (bioreactors) method considered to be the most cost efficient process for producing EPNs. Hence, this technology was the method of choice for several companies to produce multiple EPN products in many industrialized countries. Thus, account of bulk EPN product in the world market was produced by *in vitro* liquid culture technology. Presently, the majority of nematodes (including *S. kushidai, S. glaseri, S. scapterisci, S. carpocapsae, S. feltiae, S. riobrave, H. bacteriophora* and *H. megidis*) are produced in liquid culture. Among several companies, BASF (www.beckerunderwood.com), e-nema Gmbh (www.e-nema.de) in Germany, Koppert B.V. (www.koppert.nl) in the Netherlands and Certis (www.certisusa.com) in the United States are few examples, in which they are leading in liquid culture nematode production (Ehlers and Shapiro-Ilan, 2005).

Formulation and field application of Entomopathogenic Nematodes

Formulations and application methods are the obligatory fundamentals required for EPNs to consider as bio-insecticides. Formulation is a combination of (inert) substances. Formulation is anticipated to improve activity, absorption, delivery, easy to apply, efficient storage stability of an active ingredient. The concept of nematode formulation is identical to traditional pesticide formulations. But, nematodes in formulations has unique challenges like high moisture and oxygen requirements, highly sensitive to temperature extremes and behaviour of infective juveniles limit the choice of the formulation method and ingredients. Therefore, major goals of developing formulations were focusing on above mentioned critical factors including quality maintenance, enhancement of storage stability, reduction of transport costs, improvement in ease-of-transport and enhancement of nematode survival during and after application. Hence, over past several years many researchers made some experimentation on formulations.

In USA, cottage industry developed polyether-polyurethane sponge-based formulations, cadaver formulations, vermiculite formulation, alginate and flowable gel formulations. However, these formulations couldn't be successful in commercial markets due to inconsistent storage, high cost, lack of room temperature stability and difficulties in application. To overcome inconsistent storage ability, Grewal (1998) developed an improved wettable powder (WP) formulation by inducing partially desiccated nematodes through addition of water absorbents. This formulation enables storage of heterorhabditids and steinernematids at room temperature. In addition, Capinera and Hibbard, (1987) developed a granular type of formulation. In this formulation IJs were partially encapsulated in lucerne meal and wheat flour. Later, Connick *et al.* (1993) invented a 'Pesta' formulation by using wheat gluten matrix, humectant and a filler to enhance nematode survival. Further, improvement over this formulation, a water dispersible granular (WG) formulation has been developed. In this formulation, IJs were coated in 10–20 mm diam. granules consisting of a mixture of silica, clays, cellulose, lignin and starches (Georgis *et al.*, 1995). These granules were prepared through a conventional pan granulation process, in which droplets containing a thick nematode suspension were sprayed onto pre-mixed powder formulation on a tilted rotating pan (Grewal and Georgis, 1998). Granules start to form when nematode droplets come in contact with the powder formulation. Later, granules roll over the dry powder to adsorb more powder around them. At the end granules were sieved out of the powder and packed into shipping cartons.

In India, ICAR-NBAIR, Bengaluru developed a wettable powder formulations of *H. indica* and commercialized for management of white grubs and other soil insect pests. This formulation was found highly effective against soil insect pests, white grubs in particular belonging to Scarabaeidae on a number of crops such as arecanut, banana, sugarcane, potato, corn, onion etc. A gel-based formulation of an indigenous heat-tolerant EPN species, *Steinernema thermophilum* (Pusa NemaGel) was developed by Indian Agricultural Research Institute, New Delhi. This formulation was found to be effective against various insect pests *viz.*, white grubs, tobacco caterpillar, gram pod borer, brinjal shoot borer, diamond back moth, subterranean termites, etc. It can be used as soil application or foliar application depending upon the target pests.

Application of entomopathogenic nematodes

Guidelines for field application

1. Soil temperature at the time of application (and preferably even after application), should range from 18-30°C, which varies with EPN species/ strain.

2. Optimum soil moisture should be maintained at both pre and post application of EPNs.

3. Application should be done preferably in the morning or evening to avoid exposure of EPNs to UV radiation and high temperature.

4. Identification of accurate EPN species/strain/dose and susceptible life stage of the target pest is paramount importance.

5. Repetition of application is required to achieve higher percentage reduction in insect pest population.

6. About four-five week duration of trial is needed normally. For lepidopteran insects, week duration is sufficient.

7. Flood jet nozzle is fairly effective for spray.

Methods of application

1. Soil drenching (Fluids or WP formulations of EPNs): spraying nematode directly on to the soil surface is the most commonly used application method.

2. Spraying: As with chemical insecticides, spraying nematode directly on foliage is the most commonly used foliar spray method.

3. Trickle irrigation: Button type of trickle irrigation outlet is used to deliver the nematodes to the area on one side of the plant.

4. Capsule: Capsule prepared from wheat bran (5% w/w) with calcium alginate containing 1,000-2,000 nematode/capsule is applied. Capsules are buried in soil.

5. Pellet Baits: Capinera *et al.* (1987) used wheat bait pellets from lucerne meal and wheat flour (50% each), locust bean gum (18mg/ml of water) and corn oil.

6. Nylon pack cloth bands: Nematodes can be applied to nylon pack cloth bands lined with fleece or terrycloth that is wrapped around tree trunks to control gypsy moth larvae, *Lymontria dispar.*

7. Punch and Syringe method: This is used in case of forest trees. Inoculation hole is made by hammer and delivery of about 1 ml of nematode containing medium (aerated 12% gelatine with 4,000 nematodes/ml) is done by syringe.

Note: EPN application volumes/dose vary with soil type, crop, target insect, target insect life stage, target insect behaviour, formulation and plant architecture.

References

• Capinera, J.L. and Hibbard, B.E., 1987. Bait formulations of chemical and microbial insecticides for suppression of crop feeding grasshoppers. *Journal of Agricultural Entomology*, **4**:337–344.

• Connick, W.J. Jr., Nickle, W.R. and Vinyard, B.J., 1993. 'Pesta': new granular formulations for *Steinernema carpocapsae. Journal of Nematology*, **25**:198–203.

• Ehlers, R.U. and Shapiro-Ilan, D.I., 2005. Mass production. *In*: *Nematodes as Biocontrol Agents*. Grewal, P.S., Ehlers, R.U. and Shapiro-Ilan, D.I. (Eds.), CABI Publishing, Croydon, U.K. pp. 65–78.

• Georgis, R., Dunlop, D.B. and Grewal, P.S., 1995. Formulation of entomopathogenic nematodes. *In*: *Bio-rational Pest Control Agents: Formulation and Delivery*. Hall, F.R. and Barry, J.W. (Eds.), *American Chemical Society*, Bethesda, Maryland, pp. 197–205.

• Grewal, P.S. and Georgis, R., 1988. Entomopathogenic nematodes. *In*: *Methods in Biotechnology*. Hall, F.R, and Menn, J.J., (Eds.), Vol. 5, Biopesticides: use and delivery, Totowa (NJ): Humana Press Inc. pp. 271–297.

• Shapiro-Ilan, D. I., Lewis, E. E., Behle, R. W. and McGuire, M. R., 2001. Formulation of entomopathogenic nematode-infected-cadavers. *Journal of Invertebrate Pathology*, **78**: 17–23.

11

Commercialization and Practical Utilization of Entomopathogenic Nematodes

Rajeshwari R.[1] and S. S. Hussaini[2]
[1]University of Horticultural Sciences, Bagalkot
[2]ICAR- National Bureau of Agriculture Insect Resources, Bangalore

Introduction

Nematodes constitute one of the most important groups of organisms which inhabit the soil. They are highly diversified. More than a quarter of the world's population is infected with nematode parasites, and more than a hundred species of nematodes are parasites of humans. Despite extensive morbidity and mortality caused by these parasites, the biological mechanisms of host-parasite interactions are poorly understood.

Nematode fauna play a significant role in regulating primary production, predation, energy transfer, decomposition of organic matter, and nutrient cycling in soil ecosystems besides being insect parasitic through a bacterium complex and the latter have attained the status as potential biopesticides because of their impressive attributes. Nematodes of Heterorhabditidae and Steinernematidae are lethal obligate parasites of insects and are called entomopathogenic nematodes, in reference to their ability to quickly kill hosts. They possess specific biological and ecological features, which make their use in biological control exceptionally safe. The development of cost effective mass production technology and formulation led to the availability of nematode products worldwide comparable with standard insecticides in market. EPN have the potential to affect the diversity of native fauna in soil ecosystems even though they do not have any direct parasite/host or predator/prey relationship. Besides they have shown potential as antagonists to plant parasitic nematodes. Several greenhouse and field trials demonstrated the suppression of plant-parasitic nematodes by EPNs. Nematodes are used as model systems for studying complex biological systems in diverse scientific fields such as genetics, development, nutrition, environmental toxicology, pharmacology, and gerontology. The free-living nematodes (bacterivorous, fungivorous and omnivorous species) are much less studied than the parasitic species. Soil nematodes play a significant role in decomposition of organic matter and nutrient cycling (Glaser, 2002).

Nematodes are lower invertebrates, highly diversified group of animals on earth. They are the most abundant metazoans in the soil. They have been found to occupy all habitable aquatic and terrestrial habitats. Generally, the nematodes are free living in marine, fresh-water or in soil. A very large number of species are parasites of different kinds of plants and animals. The nematode soil fauna play a significant role in regulating primary production, predation, energy transfer, decomposition of organic matter, and nutrient cycling in soil ecosystems. Nematodes possess many attributes that make them useful ecological indicators for assessing and monitoring changes in the soil health.

Emphasis on biological alternatives to pesticides has increased in agriculture due to concern about environmental pollution. Ubiquitously distributed and due to lack of pathogenicity to mammals and other non-targets (NTOs), they are exempted from registration requirements. There are more than 3,100 natural associations between insects and nematodes involving 11 orders of nematodes and 19 orders of insects. The association may range from a phoretic relationship to obligate entomoparasitism and include host death, sterility, reduced fecundity, delayed development or aberrant behaviour. Some that are associated with one host and its special ecology are highly specialized and difficult to propagate on an artificial medium (some terrestrial mermithids). Other less specialized forms have wide host ranges and can be mass-produced on artificial media and are used at the present time for control of agricultural pests. There are now about 20 companies that have been successful in introducing nematode-based products into some commercial and consumer markets (Henderson et al., 2009).

Efficacy of Entomopathogenic Nematodes

The success of EPN is largely due to the extensive amount of scientific research conducted, both in the lab and field. Biological pesticide must be able to compete in relative terms with chemical pesticide in field efficacy, cost and application methods. The development of cost effective mass production technology led to the availability of nematode products comparable with standard insecticides in market. Many factors affect the ability to place quantities of nematodes on or in close proximity to the target host to produce optimal results at the lowest possible cost. To overcome the impact of abiotic and biotic factors on nematode efficacy and persistence, the inundative application of high concentration of a specific nematode species has been used as a primary control strategy. Application of nematodes is mostly targeted to the soil and against cryptic habitats presently. As with chemical insecticides, spraying nematodes directly onto the soil surface is the most commonly used application method. This broadcast method provides good coverage and is simple and quick. Local production of nematodes using *in vivo* methods provide "fresh" biologicals with improved performance. This can be taken up by the existing cottage industry.

Conservation of natural populations in agro ecosystems or the fate of applications has received little attention. Conservation biological methods are designed to protect and maintain natural or introduced enemies of pests. With EPN, these can include application practices for nematodes that will favour survival and establishment of introduced individuals besides involving crop management practices with a positive effect on existing natural or applied populations.

Expanded use of EPN in biological control cannot be expected unless field efficacy is increased. Matching nematode species might bridge the efficacy gap between the nematodes and chemicals and strains against those insects they are best adapted. Attention should be given to protecting the genetic variability of new isolates and preventing the loss of alleles through laboratory adaptation. Prediction model may have to be developed so that nematodes will be used when and where they are likely to be effective. Distinction to be made between 'lab adapted' and 'field adapted' populations.

All nematodes that emerge from a single cadaver are not necessarily the same with respect to behaviour. All the IJs are not infectious and that, at most only 10-20%

are infectious at any one time. An increase in this rate to 90% would greatly increase the effectiveness of nematode applications. Nictating behaviour is dominant in young juvenile nematodes and as they age, their lipid content falls and they change from ambushing to hunting behavior. When *S. carpocapsae* was applied to insects such as weevils which are difficult to kill, there was a population reduction of 20%. Four weeks later, there was a population reduction of 60%. It may not be advantageous for all nematodes to behave in the same way. It would be disadvantageous for all nematodes to become infective at the same time, however desirable this might be from the commercial point of view.

Matching nematodes to target hosts

In the beginning, the conventional thinking among many was that *S. carpocapsae* was some sort of biological "silver bullet". Indeed, this nematode was well known to be lethal to hundreds of diverse pest species, from termites to caterpillars, fleas to black widow spiders. This nematode seemed to have a host range not unlike the organophosphates and carbamates that then dominated the insecticide scene, and this, coupled with their exemption from government registration sparked keen interest from industry.

Yet when field-tested against these same insects, the nematode worked sometimes but more often failed, usually miserably. The conventional wisdom began to grow that nematodes "don't work" when the only truth actually revealed was that nematodes don't work for that particular insect. Just as Imidacloprid is a wonderful chemical agent against many white grub species but is nearly useless against carpenter ants. *Steinernema carpocapsae* is effective against webworms but ineffective for mushroom flies yet, *S. feltiae* is an excellent match against these flies.

In the field, behavioural and environmental barriers come into play and restrict host range. A lab-derived host range should not be confused with field activity, which can be huge. But in the real world, there are barriers that can disrupt the infection process, frustrating control efforts and resulting in a far narrower spectrum of insecticidal activity. It is these barriers that require careful matching of nematode and insect.

Infection barriers are part of the host selection process for EPN consisting of 1) host-habitat finding, 2) host-finding, 3) host acceptance, and 4) host suitability.

Each step acts as a sort of biological sieve, narrowing an experimental to a field host range. If our goal as practitioners is to match target insects with the nematode species best able to parasitize it, we must understand and appreciate each step. Parasite and host must coincide in time and space. Cabbage loopers are easy for nematodes to kill in the lab but they can rarely tolerate the physical extreme characteristic of exposed foliage such as rapid desiccation, high surface temperature, and exposure to solar radiation. Most nematodes are soil adapted. The soil environment buffers them against extremes of the aboveground world.

Once in the proper habitat, IJs locate insect hosts through ambushing and cruising strategies. If ambushing is a stationary behaviour that occurs at or near the soil surface, then it follows that ambusher nematodes are best adapted to parasitize highly mobile, surface-adapted hosts such as cutworms and armyworms. Thus, understanding host-finding strategies increases our ability to make efficacy predictions, thereby optimizing host-parasite matches. An EPN can parasitize only a single host, so each IJ must carefully assess an insect before committing irreversibly. Nematodes must be able to recognize their hosts so they don't make an irreconcilable mistake and attack a host that's unsuitable. In short, if they don't recognize a host, they shouldn't attack under most conditions. They are able to discriminate among potential hosts, the insect is not defenseless.

Production of Entomopathogenic Nematodes

In vivo, with an insect host serving as the media for nematode-bacterial growth and production requires a constant source of healthy insects, is sensitive to biological variation and costs of production are high in terms of equipment and man-hours. *In vitro* methods have been developed using monoxenic liquid culture systems that use fermentation tank technology. This approach has high initial investment in infrastructure and expertise, avoiding cross contamination.

Mass production is a key issue to EPN commercialization. Identification of potential isolate, characterization of symbiotic bacteria and suitable media are important factors. EPN possess specific biological and ecological features, which make their use in biological control exceptionally safe. Development and reproduction of EPN is impossible without the presence of their symbiotic bacteria and hence a constant source of pure cultures of primary form is essential (Ehlers, 2001).

Formulation of entomopathogenic nematodes

The first attempts at formulating EPN were initiated in 1979, but at best, shelf-life was about 1 month. IJs carried on moist substrates such as sponge, vermiculite and peat require continuous refrigeration to maintain their viability. Immobilization of nematodes in a matrix increased shelf life, but desiccation of the nematodes, which reduces energy utilization, has been the most effective means of extending their shelf-life. As more basic studies on the nematodes were conducted, distinct biochemical, behavioural and morphological differences among *Steinemematid* and *Heterorhabditid* species were documented. The relatively high lipid content of these nematodes suggests that they are adapted to survive prolonged periods of environmental stress. The selection of formulation type, ingredient, packaging size and storage conditions cannot be undertaken unless the oxygen, moisture and temperature requirements for each species are defined. Thus, a better understanding of the biochemistry, physiology and behaviour of the infective nematodes will lead to the development of formulations that are more stable and easier to use. Clearly, from moist to partially desiccated formulations points to the advances that have been made in understanding the nematodes. Nematode metabolism is temperature-dependent, with warm temperatures increasing the rate of lipid reserve used but decreasing the time that the infective juveniles remain viable and pathogenic. The successful market introduction of an EPN based-product requires a reliable and stable formulation. Although commercially formulated beneficial nematodes are used as a non-chemical alternative, they have small niche market as they do not satisfy the main criteria set up by end users. Formulation of nematode into a stable product played a significant role in commercialization. However, commercialization has been hindered by limitations in storage and application. Formulation of cadavers can increase ease of handling and storage. Application in infected cadavers would reduce production costs for *in vivo* nematode producers. They are now commercially applied in aqueous suspension. These biocontrol agents may also be applied in nematode-infected insect cadavers, but this approach may entail problems in storage and ease of handling. (Georgis, 1990)

Of 19 formulations tested (including combinations of starches, flours, clays, etc.), starch-clay combination was found to adhere to the cadaver and to have no deleterious effects on nematode reproduction and infectivity. Other formulations exhibited poor adhesion or reduced nematode reproduction. Two formulations enabled cadavers to be partially desiccated without affecting reproduction however, other formulations and non-formulated cadavers exhibited reduced reproduction upon desiccation. Four-day-old cadavers were more amenable to desiccation than 8-day-old cadavers. Formulated cadavers were more resistant to rupturing and sticking together during agitation than non-formulated cadavers. A better understanding of the biochemistry, physiology and behaviour of the IJs is a prerequisite in the selection of formulation type, ingredient, packaging size and storage conditions in development of formulations that are more stable and easier to use (Grewal, 2002).

Application technology

Strategies must also be developed which ensure the successful delivery of the nematode to the target site and target insect, thereby increasing the probability of nematode-insect interaction. Compatibility with a wide range of agrochemicals has been demonstrated. This has benefited the successful introduction with existing IPM.

Commercialization

Commercialization of EPN has experienced an assortment of highs and lows throughout the world. For every success, there have been numerous failures. In many cases, success was not achieved despite the pests having shown promising susceptibility in lab and field trials.

Products and Registration

Indigenous EPN are not required to be subjected to any kind of registration. Currently most countries do not require any regulation of EPN for biological control. In some countries, efficacy data is required and in a few others, registration is necessary. In order to harmonize the regulatory practice, it is recommended that the use of indigenous EPN does not need and should not be subjected to any kind of regulations. The release of exotic species however should be regulated.

Status of World market for EPN *vis a vis* other biopesticides

The first report on this was published by Shapiro Ilan and Grewal (2008). The production of EPN was fluctuating between 2003 and 2009 whereas the total production of biopesticides showed an increase from 195 million dollars to 257 million. The Biopesticide market is projected to reach 6.6 billion dollars by 2020 (18.8 % rate per annum.). The EPN market is expected to increase 25% by 2020. To reach this, the cost of EPN has to decrease further.

Strategies to make EPN cost effective

Factors contributing to cost are inputs, services, fees, taxes, IPR costs, building, machinery, vehicles, depending on scale of operation etc. Among the different stages, transformation, production, post production processing, formulation, packaging, transport, storage and delivery systems are critical events that cost.

Quality is regarded as a major impediment to wider adoption

Effect on nutrient cycling

Studies have indicated that EPN have the potential to affect the diversity of native fauna in soil ecosystems even though they do not have any direct parasite/host or predator/prey relationship. EPN are applied often as inundative strategy and repeated applications to control recurring pest populations may sustain the impact. Metabolic products of symbiotic bacteria of EPN are reported to possess a broad spectrum of biological activities and fundamental questions arise about their impact on soil fauna and flora and consequently affecting soil processes

EPN genetics and molecular biology

EPN belong to the same family as *Coenorhabditis elegans*, whose genome has been fully sequenced and annotated. In principle, the molecular tools which have been developed for *C. elegans* could be developed and applied to studies on EPN but, in practice, such technology transfer has been rare (Grewal and Georgis, 1988).

Photorhabdus a model organism

The complete genome sequence of *P. luminescens* subspecies *laumondii* strain TT01, a symbiont of *H. bacteriophora* isolated on Trinidad and Tobago, has been deciphered which constitutes 5,688,987 base pairs containing 4,839 protein coding genes, 157 pseudogenes, seven complete sets of rRNA operons and 85 tRNA genes. The protein coding genes are predicted to encode a large number of adhesions, toxins, hemolysins, proteases and lipase and contain a wide array of antibiotic synthesizing genes. These proteins are likely to play a role in the elimination of competitors, host colonization, invasion of insect host and bioconversion of the cadaver, thus making *Photorhabdus* a promising model for the study of symbiosis and host-pathogen interactions (Buecher and Popel, 1989).

Comparative edge over other bioagents

Growing demand for organically cultivated fresh and processed crops and produce need no stress and the current consumption of organically produced fruits and vegetables globally is valued at US$ 27 billion. Increasing exports warrant enormous efforts to produce and use biopesticides in the context of IPM. There is a greater need to adopt an integrated approach to manage pests, diseases and weeds through environment friendly and economically viable procedures of plant protection. Biocontrol agents/biopesticides have been a boon to primitive agriculture and have been economically successful in modern progressive and intensive agriculture. It is being used harmoniously with pesticide application. Biopesticides are likely to have a greater impact on the insecticides sector. Some analysts believe that biopesticides will account for 15% of the total insecticides market by the year 2010. In 1995, various economic forecasting services estimated the world market for pesticides at approximately $29 billion and the biopesticide share of the market around $380 million, approximately 1.3% of the total. Presently, biopesticides represent approximately 4.5% of the world insecticide sales. The growth rate for biopesticides has been forecast at 10-15% per annum over the next ten years.

Although EPN are extraordinarily lethal to many important soil insect pests, yet safe for plants and animals. This high degree of safety means that unlike chemicals, or even *Bacillus thuringiensis*, nematode applications do not require masks or other safety equipment and re-entry time, residues, groundwater contamination, chemical trespass and pollinators are not issues. Most biologicals require days or weeks to kill yet nematodes, working with their symbiotic bacteria, kill insects in 24-48 hours. They are highly virulent, possess chemoreceptors to search out their hosts and can be cultured easily under *in vivo* and *in vitro*. Nematode production is easily accomplished for many species. They are safe to vertebrates, plants and non-targets, have been exempt from pesticide registration in the USA, and are easily applied using standard application equipment, are compatible with many chemical pesticides and are amenable to genetic selection. Nematodes are compatible with standard agrochemical equipment including pressurized, mist, electrostatic, fan, and aerial sprayers. Application through irrigation systems has improved grower acceptance. Insecticidal nematodes are virtually without competition from other biological agents for control of soil inhabiting and plant boring insects and cost of application equals that of chemicals (Poinar, 1990).

Status of EPN research and commercialization

Field demonstration of EPN has dominated in the western countries covering thousands of hectares of land with particular reference to Florida citrus industry. Many pests of horticulture, agriculture, home and garden besides public health are controlled and EPN have reached the second position after *B. thuringiensis* based products. In India, the erstwhile Project Directorate of Biological Control, Bangalore, undertook extensive and systematic surveys for determining the existence of several potential native isolates of *Steinernema* and *Heterorhabditis*. Significant amount of EPN research on selection of improved strains, mass production and developing their simple ready-to-use formulations was rigorously pursued at erstwhile PDBC, now ICAR-NBAIR (National Bureau of Agriculture Insect Resources), Bangalore from 1996. Indian Agriculture Research Institute (IARI), New Delhi is involved in the biosystematics of EPN. Several other institutes like the Sugarcane Breeding Institute, Coimbatore, Indian Institute of Pulses Research, Kanpur, University of Agriculture and Technology, Udaipur, different State Agricultural Universities, Pulses and Rice Directorates, Cotton Research Institute, Spices Board, TERI, New Delhi are also involved in research on EPN and the list is fast expanding.

In the world, advancements have been made in commercialization to cater to most niche markets. Large numbers of companies are in the commercial production in USA and Europe. At present *S. carpocapsae*, *S. feltiae*, *S. glaseri* and *S. riobrave* can be consistently and efficiently produced with high yields per unit. *Steinernema carpocapsae* is being used against large number of pests belonging to Coleoptera, Lepidoptera and Diptera those of which are white grubs, mole crickets, fungus gnats, root weevils, bugs and borers on variety of crop plants and turfs. In UK, black vine weevil and strawberry root weevil are being controlled by *S. carpocapsae*, *S. riobrave* and *Heterorhabditis* sp. Four nematode species viz., *S. carpocapsae*, *S. riobrave*, *S. feltiae* and *H. bacteriophora* are nematode industries success stories. Differences in the field efficacy for the two EPN, *Steinernema* and *Heterorhabditis* were validated. *Steinernema* was found to be more effective against mole crickets whereas, the latter was found very effective against white grubs. Apart from insect pests, other organisms causing diseases to plants, such as plant parasitic nematodes and mites are being controlled by EPN. Besides, they hold promising against control of household pests, sucking pests and those of public and veterinary significance (Friedman, 1990).

Till date, a few companies have taken up production of EPN using *in vivo*. Most of commercial producers are located in the western world. The ease of use of nematode products is constantly being improved through formulation research. The quality of the nematodes that survive the rigors of the manufacturing process is analyzed by determining their shelf life and virulence. Nematode's shelf life is predicted from storage of energy lipid reserves of the infective juvenile nematodes and virulence is being assessed using insect bioassays. Indigenous EPN should not be subject to any kind of registration. Currently most countries do not require any regulation of EPN for biological control. In some countries, efficacy data is required and in a few others, registration is necessary. In order to harmonize the regulatory practice, it is recommended that the use of indigenous EPN does not require and should not be subjected to any kind of regulations. The release of exotic species however should be regulated.

Effect of mulches

Mulches enhanced EPN efficacy in field. When compared with bare ground, mulches enhanced control by providing cocooning sites for codling moth larvae and an easy to treat substrate that maintains moisture and enhances nematode activity. *S. carpocapsae* and *S. feltiae* were tested against cocooned sentinel codling moth larvae in cardboard strips followed by 2 h of irrigation in plots that were covered with one of four mulches (clover, shredded paper, grass hay and wood chips). Mortality of 97 and 98% were observed in paper-mulched plots respectively, compared to 80 and 76% in bare plots. The natural habitat of EPN, the soil, is a difficult environment for persistence of any organism considering its complexity of physical, chemical and biological components. Nevertheless, EPN have been isolated throughout the world in ecosystems ranging from sub-arctic to arid, temperate to tropical climates. Nematodes can survive unfavourable environment in a dormant state that prolongs their life span so that they can withstand the rigours. They are quiescent anhydrobiotes. In the soil, several factors interact. The IJs depend upon internal energy resource for survival till a host is located. So their metabolic rates and initial levels of energy reserves determine their survival in the absence of host. Post-application biology of EPN used to manage the root weevil, *D. abbreviates* was documented. Mulch was applied once in each year to study the effects of altering the community of EPN competitors (free-living bactivorous nematodes) and antagonists (nematophagous fungi), predaceous nematodes and micro arthropods). EPN augmentation always increased the mortality of sentinel weevil larvae, the prevalence of free-living nematodes in sentinel cadavers and the prevalence of trapping nematophagous fungi. Manure mulch had variable effects on endoparasitic nematophagous fungi, but consistently decreased the prevalence of trapping fungi and increased the prevalence of EPN and the sentinel mortality (Miklasiewicz *et al.*, 2002).

Effect of volatiles

Root systems of maize varieties emit (E)-beta-caryophyllene (E beta C) in response to feeding by larvae of the beetle, *Diabrotica virgifera virgifera*. The sesquiterpene attracted EPN. By comparing beetle emergence and root damage for two maize varieties, one that emits E beta C and one that does not, it was found that root protection by *H. megidis* and *S. feltiae* was higher on the emitting variety, but this was not the case for *H. bacteriophora*. The infectivity and multiplication of *S. feltiae*, *S. glaseri* and *H. indica* in granulovirus infected *C. infuscatellus* and *C. sacchariphagus indicus* were studied. Nematodes developed and reproduced in *C. infuscatellus* and *C. sacchariphagus indicus* larvae infected with granulovirus. Multiplication of IJs were significantly higher in healthy larvae of borers than granulovirus infected larvae(Ali *et al.*, 2010).

Non-target effects

In a classical biocontrol programme using EPN, it is important to understand the impact of the introduced nematode on the population of target and non-target organisms. If found promising, further evaluation on the effect of augmentation of the exotic EPNs on the target pest population be carried out and further evaluate the impact of the indigenous species on the pest population to identify the potential ones and then decide whether the indigenous species should be conserved or augmented. Non-target effects of augmenting EPN communities in soil was assessed when raw soil from a citrus orchard was augmented with either 2,000 *S. riobrave* or *S. diaprepesi*. Fewer EPN survived, if the soil had also been treated with 2,000 *S. riobrave* 7 days earlier. Thus non-target effects of augmenting the EPN community in soil vary among EPN species and have the potential to temporarily reduce EPN numbers below the natural equilibrium density (Piedra *et al.*, 2015).

Biofumigants

Mustard, used as green manure are tilled into the soil preceding crops. This acts as bio-fumigant toxic to plant-parasitic nematodes, providing an alternative to fumigants. However, a trend toward lower rates of EPN infection in fields was found where mustard green manures were applied. Mustard (*Brassica juncea*) cultivars, differing in glucosinolate levels, disrupted the abilities of *S. carpocapsae*, *S. feltiae*, *S. glaseri*, *S. riobrave*, *H. bacteriophora*, *H. marelatus* and *H. megidis* to infect insect hosts with green manure incorporated into field soil. The negative effects developed slowly in soil. Mustard biofumigation interfered with the biocontrol of insect pests using EPNs. The EPN effect on non-target beneficial is documented. On quantification of the nematode effects on soil arthropod and surface arthropod diversity in fields, more isotomid collembolans and predatory anystid mites and gnaphosid spiders under nematode applied trees indicate direct predation or indirect trophic effects (Ricardo *et al.*, 209).

Success stories

1. Citrus

It is a good example for successful commercialization of EPN. It is a long lived perennial evergreen that is grown in an orchard cropping system. Insect pests cause substantial loss of yield and quality of fruit. Citrus root weevil, *Diaprepes abbreviatus* was reported first in 1964 in Florida and now infests 15% of Florida citrus. Adults emerge throughout the year with peak in spring, feed and oviposit on foliage, first instars drop to the ground, burrow in soil and feed on roots. The chemical carbaryl was used as an adulticide or barrier treatment against neonates but only recommendation for control of larvae is, application of EPN. *S. carpocapsae*, *S. glaseri*, *S. riobrave* and *H. bacteriophora* have been found effective but most effective was *S. riobrave* with up to 90-93% control in field. Factors contributing to efficacy were biological, economical and ecological. Choice of nematode is of prime importance in success. *S. riobrave* was found to be more virulent to larvae than eight other species of EPN tested. Persistence is another factor. Stability and favourable soil conditions like moisture, texture and aeration make citrus groves amenable to EPN recycling and persistence.

Formulation and culture methods also affect efficacy. High sand content, shady canopy have facilitated easy movement of IJs and protection from UV radiation. Nematodes were only applied in the shade canopy through drip irrigation instead of broadcast application reducing cost of application. USD 62 / ha was treatment cost as against the high value of oranges, USD 6000 gross per ha. has added greatly to the success. The chemical, carbaryl required to be avoided due to imminent toxic residues in fruits (Bullock *et al.*, 1999).

2. Cotton

S. riobrave was proved highly efficacious towards pink boll worm (PBW) but nematode could not compete economically with other control strategies. Pink bollworm is an introduced pest first appearing in 1917 in US. PBW overwinter as diapausing larvae, then pupate and emerge in summer or early spring. Factors affecting efficacy of EPN are penology, environmental factors, application methods and timing. EPN was applied through irrigation systems using diverse spray equipment including aerial application. However, PBW emerging from diapauses were highly mobile and protection in localized areas was short lived. Presently transgenic cotton is considered popular option (Seenivasan *et al.*, 2012).

3. Turf

Among the different insects affecting turf, white grubs, bill bugs and cutworms are important and are amenable to control by EPN. White grub control with EPN is studied extensively. White grubs are predisposed to nematode attack and many strains of EPN have been isolated from white grubs. Various interacting biotic and abiotic factors determine efficacy. Thickness of thatch depth, accumulation of organic matter between the soil and turf grass foliage are negatively correlated with efficacy because thatch restricts nematode downward movement. Irrigation volume and frequency and soil moisture are positively related. Nematodes are more effective in fine textured soils. Hence, the efficacy varies with different species of white grubs varying degrees of susceptibility. Use is limited due to fierce competition with chemicals. Combination of Imidachlorpid and EPN are successful (Hussaini *et al.*, 2005).

Attributes of EPN as good biocontrol agents

To be successful biocontrol agents, the EPN have to exhibit certain ecological attributes

» Mass culture of EPN at affordable cost-benefit ratio is essential for the practicable solution.

» Environmental tolerance and adaptation: especially in case of foliar sprays of EPN, the considerable environmental tolerance is an essential attribute.

» Dispersal mechanism: good dispersal after their application, passive or active, is important. Therefore, a combination of ambushers and cruisers is applied together.

» Seasonal coincidence of parasite and host: all the three, the parasite, the insect and crop life-cycles have to coincide to be successful biocontrol agents.

» Physiological acceptance by the host: the host should not have immunity against the EPN.

» Biological competitiveness: nematode should multiply in host and host environment so that repeated applications not required.

Since *Steinernema* and *Heterorhabditis* qualify most of these traits, they are generally regarded as very good biocontrol agents.

Main strongholds of EPN to be successful biocontrol agents

» Like parasitoids and predators, they have chemoreceptors and are motile.

» Like pathogens they are highly virulent, killing their host within 24 to 48 h.

» They can be cultured easily, *in vitro*.

» Have a high reproductive potential.

» Have a broad host range.

» They are safe to the plant and animal health and the environment.

» They can be easily applied using standard spray equipments.

» They have a potential to recycle in the environment.

» They are compatible with many chemical pesticides.

References

- Ali, J.G., Alborn, H.T. and Stelinski, L.L., 2010. Subterranean herbivore induced volatiles released by citrus roots upon feeding by *Diaprepes abbreviatus* recruit entomopathogenic nematodes. *J. Chem. Ecol.*, **36**(4):361–368.

- Buecher, E.J. and Popiel, I., 1989. Liquid cultures of the entomogenous nematode, *Steinernema feltiae* with its bacterium symbiont. *J. Nematol.*, **21**: 500–504.

- Bullock, R.C., Pelosi, R.R. and Killer, E.E., 1999. Management of citrus root weevils (Coleoptera: Curculionidae) on Florida citrus with soil-applied entomopathogenic nematodes (Nematoda: Rhabditida). *Florida Entomologist.* **82**:1–7.

- Campos-Herrera, Raquel, Stuart, Robin, J., Pathak, Ekta, El-Borai, Fahiem E, and Duncan, L. W., 2019. Temporal patterns of entomopathogenic nematodes in Florida citrus orchards: Evidence of natural regulation by microorganisms and nematode competitors. *Soil Biology and Biochemistry*, **128**: pp 193.

- Ehlers, R. U., 2001. Mass production of entomopathogenic nematodes for plant protection. *Appl Microbiol. Biotechnol.*, **56**: 623–633.

- Friedman, M.J., 1990. *Commercial production and development*, Edited by: Gaugler, R and Kaya, H. 153–172. Boca Raton, FL: CRC Press.

- Georgis, R., 1990. Formulation and application technology. In *Entomopathogenic nematodes in biological control*, Edited by: Gaugler, R and Kaya, H K. 352Boca Raton, FL: CRC Press.

- Grewal, P.S. and Georgis, R., 1988. Entomopathogenic nematodes. In *Methods in biotechnology, Vol. 5, Biopesticides: use and delivery*, Edited by: Hall, F R and Menn, J J. 271–297. Totowa, NJ: Humana Press Inc.

- Grewal, P.S., 2002. Formulation and application technology. In *Entomopathogenic nematology*, Edited by: Gaugler, R. 265–287. USA: CABI Publications.

- Hussaini, S.S., Nagesh, M., Dar Manzoor, H. and Rajeshwari, R., 2005. Field evaluation of entomopathogenic nematodes against white grubs (Coleoptera:Scarabaeidae) on turf grass in Srinagar. *Annals of Plant Protection Sciences,* **13** (1): 190-193.

- Miklasi ewicz, T.J., Grewal, P.S., Hoy, C.W. and Malik, V.S., 2002. Evaluation of entomopathogenic nematodes for suppression of carrot weevil. *Biocontrol,* **47**: 545–561.

- Piedra-Buena, A., López-Cepero, J. and Campos-Herrera, R., 2015. Entomopathogenic Nematode Production and Application: Regulation, Ecological Impact and Non–target Effects. In: Campos-Herrera R. (eds) Nematode Pathogenesis of Insects and Other Pests. Sustainability in Plant and Crop Protection.

- Poinar, G.O. Jr., 1990. Biology and taxonomy of Steinernematidae and Heterorhabditidae". In *Entomopathogenic nematodes in biological control*, Edited by: Gaugler, R and Kaya, H K. 23–61. Boca Raton, FL: CRC Press.

- Ricardo, A.R., Donna, R., Henderson, E., Katerini, R., Lawrence, A. L. and William, E.S., 2009. Harmful effects of mustard bio-fumigants on entomopathogenic nematodes. *Biological Control*, 48 (2): *147-154.*

- Seenivasan, N. ,Prabhu, S. Makesh, S., Sivakumar, M., 2012. Natural occurrence of entomopathogenic nematode species (Rhabditida: Steinernematidae and Heterorhabditidae) in cotton fields of Tamil Nadu, India. *Journal of Natural History*, **46** (45):156-159.

12

Entomopathogenic Fungi: Opportunities and Limitations in Horticultural Crops

G. K. Ramegowda

University of Horticultural Sciences, Bagalkot

The human population is growing day by day and it is estimated that by 2030 it will reach 30 billion putting huge pressure on the natural resources especially from the food, feed and fibre point of view. At this juncture the share of global human population, potable water and arable land will be 17, 4.2 and 2.4 percent, respectively (WGR, 2016). In this perspective, our immediate challenges are we need to produce more from less land to feed more mouths. We are facing a drastic change in the climate which is not only directly affecting crop performance but the reaction of pests and diseases too. We have already exploited the natural resources and there is a need to revive and sustain them for the future too. Productivity of agriculture systems have reached the maximum and are flattened for almost a decade besides there is a huge prevalence of malnutrition. There is an urgent need to produce more for securing food, nutrition, fuel and livelihood security. Focus is now shifting towards horticulture segment to achieve these targets. From one hectare of land as many as more than five people need to be fed as against only two persons during 1960s (Tiecher, 2017a). Today, the Indian horticulture is characterised by dominance of vegetables (59.1%), fruits (31.2%) and plantation crops (5.8%) and rest is shared by spices, flowers, medicinal and aromatic crops (MOF&FW, 2018). Mushroom and honey production are also gaining importance in the recent days. Green revolution has lead to a significant increase in the Indian agriculture with four fold increase in cereals production, six times in milk, nine times in fish and a laudable 10 folds in horticulture segment. This is attributable to hybrids, high yielding varieties, quality seeds and planting materials; improvised formulations, techniques and methods of INM and IDM; plasticulture, drip and fertigation techniques offering high water and nutrient use efficiency; high density cropping and efficient crop geometries, mechanization, precision farming, post harvest technologies, marketing and warehouse networks and effective dissemination of technical knowhow from time to time. At the same time continuous increase for agriculture production has lead to increased intensification besides higher use, overuse and misuse of natural resources and build-up of pests and diseases affecting the expected targets which has escalated the use of mitigating measures with much focus on harmful synthetic pesticides (Ambethgar, 2009; Choudhary and Singh, 2017). This is further complicated by declining productivity, resistance and resurgence of pests and diseases besides residues of harmful pesticides. Uncertainty of rainfall and frequency of drought has thrown the production system out of gear. Above all the depletion of ground water has added fear to the future uncertainty. Monoculture, relay cropping and dense cropping with hybrids and high yielding varieties with poor or no host plant resistance coupled with excessive use of fertilizers and water under low organic matter system has increased the pest and disease build-up. Excessive use and misuse of these pesticides has led to a number of problems like increased protection/ production costs, resistance in the target group, elimination or threat to non-target groups, harmful and persistent residues in soil, water, food, feed and biomagnifications, etc (Shetty, 2004, Patil *et al.*, 2014). These are due to poor coverage of pesticide sprays, overdose or sub-lethal doses besides spurious standards of pesticides.

To overcome all these, rational thinking is important by involving more and more of natural pest regulating agents which are safe and sustainable in nature. Even though the deployment of natural enemies in arthropod pest management is over a century old, got momentum during 1950s and 60s. Among the various natural pest management options, entomopathogenic fungi are frequent natural enemies of arthropods worldwide (Feng *et al.*, 1994; WGR, 2016; GMI, 2017; Tiecher, 2017c). Research on these fungi has focused predominantly on aspects related to population regulation and management of arthropod pests. There are few innate features of the entomopathogenic fungi (EPF) which made their commercial exploitation progression on a slower tract (Inglis *et al.*, 2001; Jaronski, 2010; Tiecher, 2017b). They include: relatively slow action of EPFs with a minimum of one week's gestation period allowing pest damage to continue; their efficacy is largely governed by environmental conditions rather than the innate potential of EPF besides the timing of application during a day and in the season governs a lot; their efficacy can't be compared with chemical pesticides; narrow spectrum of action making a mismatch between more number of pests and EPFs; short viability of the mass produced EPFs in storage and in field and lack of economical and efficient production technologies for a great many number of EPF agents. Besides all these, lack of knowledge and suspicion about the performance of biopesticides especially EPFs besides lack of simple illustrations in IPM. Above all the only few commercial formulations are available for use have inconsistent and unplanned demands planning a production cycle (Devanur, 2010; Tiecher, 2017b; 2017c). Distortion of information during communication is another bottleneck in popularising the biopesticides in general and EPFs in specific which is another cause for non-adoption over large areas. Biopesticides production in India is largely undertaken by small national companies or entrepreneurs with a limited scale of investment with smaller production capacities making their poor/non availability in time and space. Besides all these, the investment on biopesticides is largely limited due to lack of proprietary rights with the biocontrol agents. Absence of validated standard formulation and application technologies coupled with poor/ lack of quality testing standards and labs preventing their easy marketability. Due to lack of investment and volume support, lack of greater profit margins, credit facilities, incentives in comparison to synthetic chemicals the marketing network is not inclined to promote biopesticides.

With all these in the background, huge efforts were made by ICAR, DBT, Farm Varsities with support from central and state governments of India to strengthen the biopesticide segment to provide it an industrial outlook (PPQS, 2017). They include: lowering the production cost through refinement of efficient production protocols, creating a nodal agency under DBT-DPPQS to monitor the quality of biopesticides besides creating an awareness about the use and advantages of biopesticides; simplification of registration procedures for biopesticides, incentives to farmers using biopesticides through organic certification, mechanization of production facilities to improve the quality and to reduce the cost, massive and subsidised support for production facilities to government and individual organisations, *etc.* by 2000 has gained further momentum to biopesticides which has grown from 0.2 percent of Indian pesticide market in 2000 to 4.2 percent in 2015. Global biopesticides market is mere 1.5 percent of the pesticide market and dominated by microbials with more than fifty percent share. There is a huge potential available in India to explore the infinite potential of biopesticides especially EPFs. There is a massive research infrastructure with as many as 13.23% global publications and 3.55% patents of total global holdings in the biopesticide science sector; a very constructive public support and policies with increasing consumer demand for organic products with 1.24 million tons of organic production during 2015 and of which around 0.2 million metric tons has been exported earning a good foreign exchange. Adoption of new farming practices coupled with government support for promoting bio control products through subsidies, stringent quality control protocols, demonstrations, *etc.* Now science has proved beyond doubt that the EPFs are environmentally safe due to high host specificity; have natural capability to cause epizootics due to their persistent presence in soil; compatible with all/most tactics of pest management; need lower development costs (Rs. 5 crore) than chemical agents (Rs. 500 crore); offer a long term regulation of pest suppression due to co-habitation; delayed or no resistance development in target pests; non-toxic to non-target organisms due to host specificity and hence no threat biodiversity (BBCR, 2012; Hasan, 2014).

In the Indian pesticide market which has over 227 registered products only 15 are biopesticides. Of these 15 registered biopesticides from over 150 producers across India comprise only *Lecanicillium* (13), *Beauveria* (10), *Metarhizium* (7), *Hirsutella* (1) and *Paecilomyces* (1) (Agarwal, 1990; Sinha and Biswas, 2009; Ambethgar, 2009; Ramanujam *et al.*, 2014; PPQS, 2017; Sharma *et al.*, 2018). Maharashtra, Kerala, Karnataka, Tamil Nadu and the north eastern states forms the major producers as well as consumers of biopesticides in India. Still the popularity and usage of biopesticides especially EPFs is low due to unorganised stakeholders, expensive registration procedures, inconsistency in availability of products across India, non-conformity to the quality standards and non-compliance to label specifications and claims. Above all, the low awareness among the greater section of farming community on the usage of biopesticides and their availability while, the availability of synthetic pesticides is very simple.

To increase the usage of biopesticides efforts are still on to utilize them against the increased crop pest problems due to climate change and agricultural intensification, integrating the biopesticides as an integral component of IPM especially for the production of high value crops under controlled conditions and encouraging marketing of such produce through organised sectors to get premium product pricing and providing subsidies and grants for investment to set up more and more production units across sectors. Government of India through its DBT and DST has set up a massive biocontrol research and production facilities in ICAR (31 labs), DBT (22) besides more than 200 such facilities under NATP in the states of Tamil Nadu and Andhra Pradesh alone. At the policy end the biopesticide registration process has been simplified for easier and speedier development under Insecticide act (1968) besides incorporating biopesticides under the national farmer policy-2007 for promotion of biopesticide production and utilization.

Still there are huge opportunities in Indian agriculture with special emphasis to horticulture where the consumer awareness in pushing the farming community towards residue free and/or organic pest management options in terms of preventive pest management with inundative application EFPs to fight against the insect pests along with certain diseases and plant nutrition towards which the current research is progressing. It has been proved for the sustainability in pest management with EFPs with suitable habitat/microclimate manipulation to achieve the safer pest management on a sustainable way. The focus from the industry is must to maintain the quality standards of the EPF formulations on the ethical guidelines to promote and sustain the usage and harness the benefits in an eco-friendly way.

References

- Agarwal, G. P., 1990, Entomogenous fungi in India and management of insect pests. *Indian Phytopathology*, **43** (2): 131- 142.

- Ambethgar, V., 2009, Potential of entomopathogenic fungi in insecticide resistance management (IRM): A review. *Journal of Biopesticides*, **2**(2): 177 – 193.

- BCC R, 2012. Global markets for biopesticides. http://www.bccresearch.com/pressroom/report/code/CHM029D. accessed on 10-10-2017.

- Choudhary, O. P. and Singh, D., 2017, Effect of fertilizers and pesticides use on environment and health. Paper presented in 57th Annual Conference of National Academy of Medical Sciences, India, NAMSCON 2017, 27th to 29th October 2017, Sri Guru Ram Das University of Health Sciences, Sri Amritsar, Punjab, India.

- Devanur, V., 2010, Biopesticides and emerging busisness opportunity. Paper presented in 5th Annual Biocontrol Industry Meeting, 2010, at KKL Culture and Convention Center Lucerne, Switzerland, 25th to 26th October, 2010.

- Feng, M. G., Poprawski, T. J. and Khachatourians, G. G., 1994. Production, Formulation and Application of the Entomopathogenic Fungus *Beauveria bassiana* for Insect Control: Current Status. *Biocontrol Science and Technology,* 4: 3-34.

- Global Marketing Insights (GMI), 2017. Global Biopesticides Market Size, Application Potential, Competitive Market Share & Forecast, 2024. https://www.gminsights.com/pressrelease/biopesticides-market. accessed on 10-10-2017.

- Hasan, S., 2014. Entomopathogenic Fungi as Potent Agents of Biological Control. *International Journal of Engineering and Technical Research,* 2(3): 234-237.

- Inglis, G. D., Goettel, M. S., Butt, T. M. and Strasser, H., 2001. Use of hyphomycetous fungi for managing insect pests. In: Butt, T. M., Jackson, C. and Magan, N. (Eds.), *Fungi as Biocontrol Agents- Progress, Problems and Potential.* CABI Publishing, Oxfordshire, pp. 23–69.

- Jaronski, S.T., 2010. Ecological factors in the inundative use of fungal entomopathogens. *Biocontrol,* 55: 159–185.

- Ministry of Agriculture & Farmers' Welfare (MOA&FW), 2018. Horticultural Statistics at a Glance-2018. Horticulture Statistics Division, Department of Agriculture, Cooperation & Farmers' Welfare, MOA&FW, Government of India, 458 p.

- Patil, S. K., Reddy, B. S. and Ramappa, K. B., 2014. Excessive use of fertilizers and plant protection chemicals in paddy and its economic impact in Tungabhadra Project Command area of Karnataka, India. *Ecology, Environment and Conservation,* 20 (1): 297-302.

- PPQS, 2017. Guidelines for registration. http://ppqs.gov.in/divisions/cib-rc/guidelines. accessed on 10-10-2017.

- Ramanujam, B., Rangeshwaran, R., Sivakmar, G., Mohan, M. and Yandigeri, M. S., 2014. Management of Insect Pests by Microorganisms. *Proceedings of Indian National Science Academy,* 80: 455-471.

- Sharma, K. R., Raju, S.V.S., Jaiswal, D. K. and Thakur, S., 2018. Biopesticides: an effective tool for insect pest management and current scenario in India. *Indian Journal of Agriculture and Allied Sciences,* 4 (2): 59-62.

- Shetty, P. K., 2004. Socio-Ecological Implications of Pesticide Use in India, *Economic and Political Weekly*, **39** (49): 5261-5267.

- Sinha B. and Biswas, I., 2009 Potential of Biopesticide in Indian Agriculture vis-a-vis Rural Development, India, In: *Banarjee, P. (Ed.), Science and Technology-2008*, NISTADS, CSIR, New Delhi, pp. 340-342.

- Tiecher, H., 2017a. Biopesticides, Part I: Bridging the Conventional Interface. https://medium.com/@harald.teicher/biopesticides-part-i-bridging-the-conventional-interface-3aad6fdf0539. accessed on 10-10-2017.

- Tiecher, H., 2017b. Biopesticides, Part II: A Definition of Success. http://biocomm.eu/2017/09/05/biopesticides-part-ii-definition-success/. accessed on 10-10-2017.

- Tiecher, H., 2017c. Biopesticides, Part IV: Market Drivers and Trends. https://www.linkedin.com/pulse/biopesticides-part-iv-market-drivers-trends-bioscience-solutions/ accessed on 10-10-2017

- Wise Guy Reports (WGR), 2016. Bio Control Agents - Global Market Outlook (2015-2022), WGR456516, 127 p. https://www.wiseguyreports.com/reports/456516-biocontrol-agents-global-market-outlook-2015-2022. accessed on 10-10-2017.

13

Entomopathogenic Fungi: A Biological Weapon for Pest Management

Prasad Kumar and Rajeshwari R.
University of Horticultural Sciences, Bagalkot

Introduction

Insects, like other organisms, are susceptible to variety of diseases caused by virus, bacteria, fungi, protozoa, rickettsia, mycoplasma and nematodes. Some of these pathogens may be quite common and frequently cause epizootics in natural insect populations. Insect diseases and their symptoms have been recognized as far back as 2700 BC in China with honey bee, *Apis mellifera* and Silk worm, *Bombyx mori*. Ancient Indian literature also refers to diseases of these insects. So far over 3000 micro-organisms have been reported to cause diseases in insects. Leconte (1873) for the first time advocated the use of diseases as means of insect control.

A disease can be brought about in a susceptible host by the pathogen through the effects of chemical or toxic substances, the mechanical destruction of cells and tissues and the combination of these two actions. There are two general types of toxins produced by entomopathogenic organisms ie. Catabolic and Anabolic substances. The catabolic toxins results from decomposition brought about by the activity of pathogen. They may arise from the host substrate or from the decomposition of pathogen itself. For example, the breakdown of proteins, carbohydrates and lipids by the pathogen may produce toxic alcohols, acids, alkaloids etc. Anabolic toxins are synthesized by the pathogen and these may be classified as exotoxins and endotoxins. The exotoxins are excreted or passed out of the cell of the pathogen. Bacteria and fungi are known to produce exotoxins. The endotoxins, produced by the pathogen are confined to the cell and are liberated when the pathogen dies or degenerates.

The first pathogen found to cause disease in insects were fungi because of their conspicuous macroscopic growth on the surface of the host. The disease caused by fungi are termed as mycoses. Over 750 fungal species are known to attack terrestrial and aquatic arthropods. Entomopathogenic fungi are found in the division Eumycota and in the sub divisions; Mastigomycotina, Deuteromycotina, Zygomycotina, Ascomycotina and Basidiomycotina. The development of fungal infections in terrestrial insects is largely influenced by environmental conditions. High humidity is vital for germination of fungal spores and transmission of the pathogen from one insect to another (Ambethgar, V., 2009).

The green mascardine fungus, *Metarhizium anisopliae* has been detected from atleast 300 species of insects. It is especially active against Chrysomelid, Curculionid and Scarabid beetles. A related species, *Metarhizium flavoviridae* has been found promising for the control of desert locust and grasshoppers. *Beauveria bassina* causes a disease known as the white muscardine. *Lecanicillium lecanii* is common pathogen of scale insects in tropical and subtropical environment. *Nomuraea rileyi* is used against army worms and loopers. *Hirsutella thompsonii* is used against mites. The effectiveness of fungi against pest depends on having the correct fungal species and strain with the susceptible insect life stage at the appropriate humidity, temperature and soil texture.

Metarhizium anisopliae

Metarhizium anisopliae, formerly known as *Entomophthora anisopliae* is an Entomopathogenic fungus that infects insects that come in contact with it. It grows naturally in soil throughout the world, causes disease in various insects by acting as a parasitoid. The first use of *Metarhizium anisopliae* as a microbial agent against insects was in 1879, when Elie Mechnikov used it in an experiment to control the wheat grain beetle, *Anisoplia austriaca*. Although *Metarhizium anisopliae* is not infectious or toxic to mammals, inhalation of spores could cause allergic reactions in sensitive individuals.

Mode of action

The disease caused by the fungus is sometimes called green muscardine disease because of the green colour of its spores. When these mitotic (asexual) spores (conidia) of the fungus come into contact with the body of an insect host, they germinate and the hyphae that emerge penetrate the cuticle, hyphal growth continues until the insect is filled with mycelia. The fungus then develops inside the body eventually killing the insect after a few days. *Metarhizium anisopliae* can release spores under low humidity conditions (<50%). However some insects have developed physiological mechanisms to reduce infection by fungi (Bhukari et al., 2011). The desert locust produces antifungal toxins, which can inhibit the germination of spores, in addition, insects can escape infection by moulting rapidly or developing a new integument before the fungus can penetrate the cuticle. It does not appear to infect humans or other animals and is considered as safe as an insecticide.

Target pest

Root weevil, plant hoppers, japanese beetle, black vine weevil, spittle bug, thrips, white grubs, Rhinoceros beetle grub, *Helicoverpa*, sugar cane pyrilla, termite etc.

Dosage

Foliar spray   : 5 g/ l of water. The spray volume depends on the crop canopy.
Soil application : 5.0kg/ acre
Drip system   : 5g/l of water.

Commercialization

The toxins produced by *M. anisopliae* are Destruxins A, B, C, D, E and F. *Metarhizium anisopliae* is grown on a large scale in semi-solid fermentation similar to that used in the production of *B. thuringiensis*. The spores can then be formulated as dust. The fungal spores can also be grown on sterilized rice in plastic bags for small scale production. *M. anisopliae* is sensitive to temperature extremes, spore viability decreases as storage temperature increases and virulence decreases at low temperatures. Bioblast is commercially available formulation that is used to control termites such as *Reticulitermes* sp. The fungus is applied into wood in to certain active termite galleries. The termites in these galleries are exposed to direct contact with the fungus.

Role in pest control

Unique among entomopathogens, fungi do not have to be ingested and can invade their hosts directly through the exoskeleton and cuticle. Therefore, they can ingest non-feeding stages such as eggs and pupae. The site of invasion is often between the mouth parts, at intersegmental folds or through spiracles, where locally high humidity promotes germination and the cuticle is non sclerotized and more easily penetrated. *Metarhizium* is an opportunistic hemitroph with a parasitic phase in the living host and saprophytic phase during post mortem on the cadaver. The fungi may use toxins to help overcome host defences.

Commercially available formulations of *Metarhizium anisopliae*

Trade name	Company
Biomagic	T. Stanes and Company Limited, Coimbatore
Metarhizium anisopliae	Multiplex, Bengaluru
Toxin (*Metarhizium anisopliae*)	Varsha bioscience and Technology, Bengaluru
Green meta	Green life Biotech lab, Coimbatore
Metarhizium anisopliae	Jay Meenakshi Bio Agro, Thirumangalam, Madhurai
Devastra 2% AS and Kalichakra 1% WP	International Panacea Limited, Haridwar

Beauveria bassiana

The origin of microbial pest control date back to the early nineteenth century, when the Italian scientist, Agostino Bassi spent more than 30 years studying white muscardine disease in silkworms (*Bombyx mori*). He identified *B. bassiana* named in his honour, as the cause of the disease. The white muscardine fungus has been detected from over 700 species of insect.

Mode of infection

The toxin produced by *B. bassiana* are Beauvericin, Beauverolids and Basssinolid.

The action of *B. bassiana* on insects begins from the penetration of spores in a body cavity through cuticle. After penetration in to the body, the spores germinate in to hyphae, then a mycelium overgrows from which conidia split off. The conidia begin to circulate in haemolymph. In this stage, the infection of insect is the consequence of the excretion of the considerable quantity of toxins. If toxin is absent, the mycelium gradually fills up the whole body of the insect. In the beginning, muscular tissue is affected. Fungus growth continues until all the tissues are destroyed. The fungus can form conidiophores, which rupture the cuticle and the envelope of a dead larva. The affected insect is covered with white conidiophores and then spore maturation and mass sporulation begins.

Target pests

Aphids, leaf hoppers, whitefly, mealybug, psyllids, chinch bug, Japanese beetle, Colorado potato beetle, cabbage caterpillar, codling moth, cotton boll worms, coffee berry borer, castor semilooper, rice brown plant hopper etc.

Commercially available formulations of *Beauveria bassiana*

Trade name	Company
Bio power, Dispel, Larvocell	Excel Industries Limited, Mumbai
Don Muscadin	Osian Agro India Private Ltd, Baroda, Gujarat
BABA	Multiplex, Bengaluru
Abtech- Beauveria	Agro Bio-tech Research Centre, Kottayam, Kerala
Boverin	NZIM Agro, Ukraine

Lecanicillium lecanii

It is formerly known as *Cephalosporum lecanii* and *Verticilium lecanii*. It was first described in 1861 and is a cosmopolitan fungus found on insects. It is an entomopathogenic fungus, the mycelium produces a cyclodepsipeptide toxin called bassianolide and other insecticidal toxins such as dipicolinic acid. It is also known as a white halo fungus because of the white mycelial growth on the edges of infected insects.

Target pests: aphids, whiteflies, leaf hoppers, mealy bugs, coffee green scale etc.

Mode of action

The conidia of *L. lecanii* are slimy and attach to the cuticle of insects. The fungus infect insects by producing hyphae from germinating spores that penetrate the insect's integument. The fungus then destroys the internal contents and the insect dies. The fungus eventually grows out through the cuticle and sporulates on the outside of the body. Infected insects appear as white to yellowish cottony patches. The insects are diseased in 7 days. However, due to environmental conditions there may be some considerable lag time from infection to death of insect. *L. lecanii* works best at temperatures of 15-25°C and relative humidity of 85-90%. The fungus requires high humidity for at least 10-12 hours. *L. lecanii* strains with small spores infect aphids; where as fungal strain with large spores infect white flies.

Commercialization

Higher doses of fungus result in faster kill. Virulence depends on the density of spores and rate of sporulation, which is dependent on environmental conditions. Fungal virulence varies with the method of conidial production. Less virulent conidia are obtained from fermented media as compared to solid media. Formulated products from conidial production can last up to one year. These products are easy to wet and dilute for spraying, also the fungus can stick to the surface of leaves and host insects. Studies have shown that combining entomopathogenic fungi with an insecticide may enhance its performance as the fungi create wounds that make it easier for the insecticide to enter the insect.

Application

Spray volume depends on crop canopy and sprayed @ 5 g/l of water. For drip system, 5 g/l of water is recommended. The resultant solution should be filtered before injecting into the drip system. Application should be repeated at least once in 15-20 days for 4 times. For green house pest problems, applications at every 10-15 days is recommended. All applications should be based on monitoring of pest populations.

Commercially available formulations of *L. lecanii*

Trade name	Company
Biocatch	Excel Industries Limited, Mumbai
Verticel	Excel Industries Limited, Mumbai
Varsha	Multiplex, Bengaluru
Abtech Verticillum	Agro Bio-tech Research Centre, Kottayam, Kerala
Vertalec and Mycotol	Koppert biological systems Pvt. Ltd., Bengaluru

Microbial insecticides can be applied in the field

1. As introduction or colonization

2. As sprays, dusts or baits

3. As mixture with insecticides

4. Employed through parasitoids and predators

Advantages of entomopathogenic fungi

» High degree of specificity for pest control

» Minimum effect on non-target organisms

» A high potential for further development by biotechnological research

» A high persistence in the environment provides long-term effects of entomopathogenic fungi on pest suppression

» Microorganisms have natural capability of causing disease at epizootic levels due to their persistence in soil and efficient transmission

» Many insect pathogens are compatible with other control methods including the chemical insecticides

» The large scale culture and application is relatively easy and inexpensive in most cases

» Resistance in insects to entomopathogens is not developed or may develop slowly as compared to insecticides

Disadvantages of entomopathogenic fungi

> » Relatively slow in action.

> » A wide range of environmental factors affect the efficacy for pest control

> » Application needs to coincide with high relative humidity, low pest numbers, and a fungicide free period

> » Due to the high specificity, additional control agents are needed for other pests.

> » Short shelf life of spores necessitates cold storage

> » The persistence and efficiency of entomopathogenic fungi in the host population varies among different insects species, thus insect-specific application techniques need to be optimized to certain long-term impacts

> » It is necessary to correctly time of the application, as most pathogens cause disease after a definite incubation period which varies in different stages of insects

> » It is necessary to maintain the virulence and viability of the pathogens till use

> » A potential risk to immunosuppressive people.

References

• Ambethgar, V., 2009, Potential of entomopathogenic fungi in insecticide resistance management (IRM): A review. Journal of Biopesticides, 2(2): 177 – 193.

• Bukhari, T., Takken, W. and Constantianus J.M.K., 2011. Development of Metarhizium anisopliae and Beauveria bassiana formulations for control of malaria mosquito larvae. Parasites and Vectors, 4: 1-14.

14

Recent Advances in *Bacillus thuringiensis* Formulations

M. C. Nagaraju, M. Mohan*, Basavarya
*ICAR-National Bureau of Agriculture Insect Resources, Bangalore

Introduction

Bacillus thuringiensis Berliner (Bt) based biopesticides are utmost importance and occupy almost 97% of the world biopesticide market. Bt is a Gram positive, spore forming bacteria that has insecticidal properties (also called "entomotoxicity") affecting a selective range of insect orders, namely, Lepidoptera, Diptera and Coleoptera. Genetic engineering may play a complementary role in the development of more efficacious formulations by facilitating greater toxin production, broadening the host range, and enhancing germination, sporulation and expanding Bt spectrum (Satinder *et al.*, 2006).

Harvesting Techniques

The final fermented Bt broth comprises of spores, cell debris, inclusion bodies, enzymes and other residual solids, which needs to be recovered efficiently. Key factors governing the choice of harvesting strategy include process throughput, physical characteristics of product and impurities and desired end-product concentration. Most commercial Bt products contain insecticidal crystal proteins (ICP), viable spores, enzyme systems (proteases; chitinases; phospholipases), vegetative insecticidal proteins and many unknown virulent factors along with inerts/adjuvants. Earlier, lactose-acetone technique was used as a method to recover Bt spores with measurable losses. However, use of advanced methods like ultracentrifugation, microfiltration and vacuum filtration to separate insoluble solids (active ingredients) from soluble liquid (inert) fraction of the harvest liquor, has resulted in efficient recovery of the active ingredient (a.i.) of various biotechnological products (Rowe et al., 2004).

Harvesting microorganisms from submerged fermentation is often difficult due to low concentration of products, their thermolabile nature and in some cases, poor stability. Stabilizing adjuvants may have to be incorporated in postharvest operation to prevent spore mortality and/or germination. Rapid drying or addition of specific biocidal chemicals may be required to prevent growth of microbial contamination in the broth or centrifuge slurry. This method could find utility only for small volumes of broth liquid. Even, foam flotation process has been applied to obtain crystal-enriched suspensions of Bt in which gelatin caused spores to be selectively entrained in the foam and thereby separated from suspensions.

Method of spray drying for large broth volumes, which can be preceded by thickening of the fermentation liquid by centrifugation and filtration using filter aids like celite, superfloc, etc., to reduce handling volume. In the spray drier, water is removed from the broth slurry as it passes through the heated inlet (150-200°C). The resulting powder coats the walls and collects in the spray dryer. An efficient recovery of active spore-crystal complex of Bt was reported using either a disk-stack centrifuge or a rotary vacuum filter with spore recovery efficiency higher than 99%. The concentration of dry solids produced by filtration (31.5%) was superior to centrifugation (7.5%) (Satinder et al., 2006).

Why formulations are required?

Formulation development can play a key role addressing four major objectives which can serve as benchmarks for success:

1. Stabilize microbial agents during distribution and storage

2. Aid handling and application of the product

3. Protect agent from adverse environmental factors

4. Enhance activity of microbial agents in field

Principally, a formulation comprises *a.i.* (fungi, bacteria, virus, nematodes, *etc.)* and additives (various substances to maintain status quo of *a.i.*) to fulfill afore stated objectives and give shape to biopesticides (Tsuji, 1997).

Adjuvants/additives

They are chemically and biologically active compounds that can alter the formulation physics and kill the targeted species without harming other insects and reduce the effective biopesticide dose required. Registration agencies like US Environmental Protection Agency (US EPA) regulate the inclusion of certain ingredients in adjuvant formulations and hence testing of adjuvant and/inerts is restrained within the lists 4A and 4B comprising minimal risk and no risk inert ingredients respectively. For example, in the past, xylene was used as a preservative in Bt formulations, but its adverse environmental impacts later led to the withdrawal of this concept.

Key factors in selection of appropriate adjuvants

Formulation type

» Solid, liquid, encapsulated forms

Target(s)

» Pest species (type)

» Developmental stage, dense or sparse growth (Low/high volumes of spray?)

» Barriers to penetration (waxy, hairy or thick leaves, sediments)

» Method of application (aerial/terrestrial; foliar spray, boom and nozzle spray, hydraulic) and timing

» Volume application rate and spray droplet size spectrum

Environment

» Site conditions (aquatic or terrestrial? In sensitive areas?)

» Current conditions (air temperature? Windy? UV radiation? Rain?)

» Water chemistry (hard or soft water? Low or high pH?)

Other(s)

>> Product interactions or compatibility issues

>> Order of incorporation into the tank mix

>> Avionics use and buffer zone width

Different adjuvant/ additives with their functions and examples (Satinder *et al.*, 2006).

1. **Dispersant:** Dispersion of formulation into dispersant medium.
   Example: Amylose, Aluminium silicate, Sodium starch glycolate etc.

2. **Surfactants and wetters**: Enhance the emulsifying, dispersing, spreading, sticking or wetting properties of the biopesticide (includes spray modifiers).
   Example: Ethoxylates (Tween/Triton series), polyethylene glycol etc.

3. **Stickers and spreaders**: Adhesion of pesticides onto the foliage protecting from rain wash-off and spreading evenly for maximum coverage.
   Example: Gelatin; gums, molasses, skimmed milk, proprietary like Nufilm and chevron, vegetable gels, vegetable oils, waxes, water-soluble polymers etc.

4. **Drift control agents/anti-evaporants/ humectants** : Reduce spray drift, which most often results when fine spray droplets are carried away from the target area by breezes, including those caused by the vehicle carrying the spray equipment and control of foam while mixing.
   Example: Polyacrylamides, polysaccharides and certain types of gums, sorbitol, sucrose, molasses, polyglycol, molasses, glycerol.

5. **Thickening agents**: Modify the viscosity of spray solutions and are used to reduce drift, particularly for aerial applications.
   Example: Water swellable polymers producing a "particulate solution" by hydroxyethyl celluloses and/or polysaccharide gums.

6. **pH Buffers**: Adjust the buffer pH; improve the dispersion or solubilisation in the formulation, control its ionic state and increase adjuvant compatibility.
   Example: Sodium phosphate, potassium phosphate etc.

7. **Defoaming and antifoam**: agents reduce surface tension, physically burst the air bubbles, and/or otherwise weaken the foam structure
   Example: Dimethopolysiloxane-based, silica, alcohol and oils.

8. **UV radiation screens:** Protect from the deleterious effect of sunlight by forming a protective layer on the formulations.
   Example: Congo Red, folic acid, lignin, molasses, p-aminobenzoic acid, alkyl phenols etc.

9. **Phagostimulants**: Stimulate feeding of formulations by pests
   Example: Corn meal, sucrose, wheat germ, corn germ, soya flour, casein, edible oil etc.

10. **Synergists**: Multiple modes of action; generally complements various formulation components.
    Example: Sorbitol, sorbic acid, sodium phosphate, stilbene, tinopal, silicate, protease inhibitors, oleic acid, linoleic acid etc.

11. **Anti-microbial agents:** Suppresses the growth of other microorganisms, retaining formulation purity.
    Example: Sorbic acid, propionic acid, crystal violet etc.

12. **Carriers** Aid in delivery of formulation to target.
    Example: Alginate, carrageenan, peat, acrylate and acrylamide supersorbents, diatomaceous earth etc.

13. **Binders:** For binding the particulates in granules together.
    Example: Gums, molasses, PVP, resins etc.

14. **Suspending agents:** Keep the formulation in suspension.
    Example: Sorbitol, soya polysaccharides, starch glycolates, sucrose etc.

15. **Attractants**: Act as baits to attract target pests.
    Example: Pheromones, cucurbitacin and various alkaloids, plastisol (PVC and cotton seed oil) etc.

16. **Multi-purpose:** Perform various functions at the same time.
Example: Molasses, starch, lignin etc.

Formulations can be classified into

1. Dry solid (dusts, granules, powders and briquettes) and

2. Liquid (termed "suspensions, oil or water based and emulsions) formulations.

Advantages of dry solid formulations

» Are ready-to-use such as dusts, granules and briquettes, applicable with simple equipment.

» Granules show less drift and applicable to hidden foliage

» WPs are easy to transport, store and apply when required, low risks of operator safety

» Cost effective; transportation costs are low

» Do not need high quantity of preservatives

Disadvantages of dry solid formulations

» Dusts are subject to drifts and pose user hazards

» Some WPs could clog sprayers

» Formulation involves harsh spray drying steps-loss of a.i.

» Use restricted to gardens, agriculture and water streams

» More expensive to apply

Advantages of liquid suspensions

» Emulsion concentrates and suspension concentrates hold high a.i., bulk storage not necessary

» ULV concentrates can be used without mixing

» Encapsulated suspensions increase residual toxicity and decrease user hazard

» Development process is without harsh conditions of drying-higher recovery

» Used in agriculture, gardens and forests

» Less expensive to apply

Disadvantages of liquid suspensions

» Subject to deterioration on long storage

» *a.i.* may settle out of emulsions and suspensions, at times

» Sometimes, require complex spray equipments, e.g., in forestry

Dry Solid Products

Dusts

Dusts are formulated by the sorption of an *a.i.* onto a finely ground, solid inert such as talc, clay or chalk, with particle size ranging from 50-100 mm. Although, finer particles adhere better, at the same time, they pose serious inhalation hazard for the user and drift hazard for the sprayer. Particle size (0.5-50 mm), bulk density (0.5–0.6 g/cm^3) and flowability are important control parameters. During application, smaller particles collect on target surfaces and large ones fall off and lack stickiness. Thus, stickers (adherents) and desiccants (prevent caking) are commonly employed. Bt dusts have been in stored bulk grains to control surface-dwelling lepidopteran pests and also Bt dusts have been widely used in the control of corn borer larvae.

Granules

The granules comprise discrete masses of 5-10 mm^3 formulated by using carriers like clay minerals, starch polymers, dry fertilizers and ground plant residues (Navon *et al.*, 1997).

Concentration of organisms in granules is 5-20%. There are three types of granules

1. Exterior granules-microbes attached to outer surface of a carrier by a sticker

2. Exterior granules without stickers

3. Incorporated granules-all constituents mixed into a paste to form matrix and later sieved to desired size. Normally, they are employed in agricultural crops, e.g., cabbage, corn etc.

Various types of granules - wheat meal granules, corn meal baits, granules formed with gelatinized corn starch or flour, casein, gluten, cottonseed flour and sugars, gelatin or acacia gum, sodium alginate and paraffin, diatomaceous earth and semolina commonly employed as formulations. In a recent advancement in granular formulations, solutions containing powder formulations of *B. thuringiensis var. israelensis* (Bti) or *B. sphaericus* were transformed into ice pellets (named IcyPearls). This technique encompassed various advantages over Bti sand granules:

1. Bti ice pellets melted on the water surface and released the microbial crystals

2. There was no loss of Bti by friction in the spraying equipment

3. The ice formulation resulted in increased swath widths, significantly reducing application costs.

Briquettes

They are large blocks with sizes ranging from 100 to 250 mm and possess same carriers as in granules presenting no drift problems. Bti based briquette formulations are largely utilized in public health sector for control of mosquitoes. Various formulations of Bti have given higher rates of kill and sustained persistence in several cases, ranging up to two months in single application. Briquettes have been normally made using organic polymers like polyvinyl alcohol to give floatability causing sustained release of toxins over several months. Mostly floating type formulations with carrier materials such as wheat flour are common with Bti and also buoyant forms which have self-encapsulating abilities and gypsum has been used to enhance sustained release of toxins (Gunasekaran *et al.*, 2002).

Wettable powders (WPs)

They consist of 50-80% technical powder, 15-45% filler, 1-10% dispersant and 3-5% surfactant by weight to achieve a desired potency formulation. Fillers are hydrophilic and usually contain silica which resists cake formation and friability during grinding. However, silica content must be kept low to avoid abrasion of formulation equipment.

Among the dried formulations of biopesticides, much attention has been paid to WPs because of their longer shelf life, good miscibility with water and ease of application as sprays with conventional equipment. Bt sp. *aizawai* based WP with 55% suspensibility, 24 secs for wetting time and 5.69×10^4 CFU/ml of LC50 value against *Spodoptera exigua* larvae was used.

Liquid Suspensions

Suspension concentrates (flowables)

They are suspensions of particulates in liquids, with 10-40% microorganism, 1-3% suspender ingredient, 1-5% dispersant, 3-8% surfactant and 35-65% carrier liquid. Based on type of carrier liquid, surfactants are characterized by hydrophile–lipophile balance (HLB) whereby lower HLB is suitable for water-in-oil and vice versa for oil-in-water based suspensions. In forestry generally, ultra-low volume (ULV) sprayers (required dispersion rate of 20 BIU/ha in a final volume of 2.5 litres/ha) are preferred to deliver higher Bt concentrations. Earlier, the suspension concentrates were ready-to-use formulations requiring just the addition of sticker (0.06%) which minimized spraying cost by eliminating mixing time and maximizing payload. Some authors had also suggested addition of sorbitol as a dispersing agent, which enhanced density of Bt formulations resulting in concentrated suspensions (reducing loading costs) and acting as an anti-freeze as well as anti-evaporant further structuring formulations. Liquid formulations of *B. sphaericus* have also been successfully employed in rice fields, ponds and other water sources against mosquitoes from the genus *Culex*.

Emulsions

They comprise of liquid droplets dispersed in another immiscible liquid (dispersed phase droplet size ranges from 0.1 to 10 mm), e.g., oil-in-water (normal emulsions) and water-in-oil (invert emulsions). The emulsions do not encounter sedimentation problems, but creaming and layer separation are common. In biopesticide jargon, they are referred to as suspo-emulsions. As oil is external phase in invert emulsions, losses due to evaporation and spray drift are minimal. However, lower shelf stability and phytotoxicity may affect the overall performance of the emulsions.

Encapsulations

They are recent advances in bioinsecticidal formulations and provide protection from extreme environmental conditions (UV radiation, rain, etc.) and enhanced residual stability due to slow release of formulations (sustained delivery). They are usually liquid suspensions with possibility of powders and granules. Microbial propagules (e.g., Bt) are encapsulated in a coating (capsule) made of gelatin, starch, cellulose and other polymers and even microbial cells.

Autoencapsulated (biological origin) formulations against European corn borer (*Ostrinia nubilalis*), were made by mixing starch powder and sugar. Fine, encapsulated products can be sprayed in any volume as the pathogen is held tightly to additives causing less wastage. Another recent advance in encapsulations is production of hydrocapsules that are of shell core type (water based), consisting of a polymer membrane surrounding a liquid centre. These shells are produced by using UV radiation initiated free-radical copolymerization of functionalized prepolymers (silicones, urethanes, epoxys, polyesters, etc.) and/or vinyl monomers such as acrylates for better dispersion and UV radiation protection

Encapsulation in the form of microcapsules has been extensively exploited to give smaller size, highly efficient fungal biopesticide formulations. This technology could be extended to Bt suspensions, which would enhance aerial dispersion onto foliage and feeding, by larvae (Guerra *et al.*, 2000).

Bacillus thuringiensis is a highly efficacious bioinsecticide used to control lepidopteran pests in the field. Unfortunately, it has limited residual activity on plants because sunlight inactivates spores and crystals and they can be washed off by rain. To minimize loss of activity, formulations must contain UV protectants, stickers, or both. Eighty formulations were tested and optimal combinations of ingredients and spray drying conditions have been determined for improving *B. thuringiensis* residual activity after simulated rain and simulated sunlight.*B. thuringiensis* stability, after simulated sunlight (xenon light/8 h) and rain (5 cm/50 min), was improved using formulations based on lignin, corn flours, or both, with up to 20% of the active ingredient, when compared with technical powder or Dipel 2x in laboratory assays (Guerra *et al.*, 2002).

Two formulations, made with corn flours and lignin 1 pregelatinized corn flour (PCF), killed 51.6 and 75.3% of *Ostrinia nubilalis* (Hübner) neonates after rain, respectively, versus 27% for technical powder. When the insecticidal activity was tested after simulated sunlight, corn flour-based formulations killed 78.5% of test larvae and the lignin 1 PCF formulation killed 70.4%, in contrast to technical powder which caused an average of 29% mortality. Formulations made with Dipel 2x rather than technical powder, caused 62.5% mortality (corn flour-based formulations) and 72.3% mortality (lignin 1 PCF), versus 53.4% for Dipel 23 after rain.

Booster formulations (enhanced entomotoxicity)

These formulations fall into a different class as there is enhancement of entomotoxicity by either stimulant or synergistic action. Addition of 675 mg/L monosodium glutamate to commercial formulation of Bt ssp. kurstaki, (DiPel12X DF), lowered LC50 from 450 to 150 mg/ L (P < 0.05, Lethal Ratio Significance Test), indicating its potential to enhance entomotoxicity and economy of Bt based formulations. Lytic enzyme like chitinase could also increase entomotoxicity by perforating the peritrophic membrane barrier in the larval midgut and thus increasing accessibility of Bt d-endotoxin molecule to its receptor on epithelial cell membranes.

Formulation of Bt toxin with nanoparticles

Bt as Nanoparticles is "Nanopesticides" involve either very small particles of pesticidal active ingredients or other small engineered structures with useful pesticidal properties. Nanopesticides can increase the dispersion and wettability of agricultural formulations (i.e., reduction in organic solvent runoff), and unwanted pesticide movement. Nanomaterials and biocomposites exhibit useful properties such as stiffness, permeability, crystallinity, thermal stability, solubility, and biodegradability needed for formulating nanopesticides. Nanopesticides also offer large specific surface area and hence increased affinity to the target. Nanopesticides, including nanoparticles, nanoemulsions have a particle size of 100-200 nm, exhibit useful properties such as stiffness, permeability, crystallinity, thermal stability, solubility and biodegradability (Swamy and Asokan, 2013).

Benefits of *Bt* as nanopesticides

The main mode of action of this crystalliferous bacterium in different orders of insects was disruption of the epithelial lining of the mid-gut. The toxicity arises when solubilized toxic fractions of the crystal attack certain receptors lining mid-gut causing perforation followed leaking of the alkaline gut contents into the hemocoel which results in changes in the pH and paralysis. The crystal (delta-endotoxin) protein must be ingested, solubilized and activated by larval gut enzymes to form its entomocidal effect. *Bacillus thuringiensis* has its specific approach to the host organism. It causes great physiological changes in the vital systems during the course of infection. It affected the total protein content and interfered with the activity of enzymes which play a dominant role in insect metabolism. Bacterial formulations were found to be not only safe for human beings and their domestic animals and caused no environmental contamination, but also they were highly specific and effective as insect control agents, besides, they could be easily prepared even in small laboratories or industrial factories.

Nanopesticides utilize and explore the ambit of Nanotechnology, which underlines the concept of particle size reduction and its properties. When the particle size comes to nano form (1nm to 100nm), the property and behaviour of matter changes due to quantum size effect. For example, when the particle of an element becomes very small (nano meter) in size, physically it facilitates more surface area to volume ratio leading to the changes in the physical, mechanical, electrical and optical properties. For instance, the increase in surface to volume ratio of some element could change the mechanical, thermal and catalytic properties of that element. By utilizing this peculiar features, Nano-encapsulation of the nanopesticide is being done to minimize the doses and to get maximum effect with more target oriented action of the pesticides (Swamy and Asokan, 2013).

The pesticide or the chemical compound is encapsulated inside on an artificially synthesized nano-matrix. In the nano-encapsulation, nano-matrix coat is synthesized artificially in such a way that it facilitates the release of the pesticide only in the targeted environment like in specific pH condition, specific temperature or in the presence of specific compounds. For example, the pesticide will be released in the intestine as the synthesized nano matrix will be dissolved in the environmental condition of the intestine of that specific insect and will kill the targeted insect only. Nanopesticides for specific insects can be made using different nanomaterials. This reduces the doses as compared to the traditional pesticide where large doses of pesticides are sprayed. Thus, it controls the release of toxic substances to the environment causing no harm to other beneficial organisms. Some encapsulation is done in such a way that it gets absorbed in the surface of the plant and facilitates prolong release which lasts for longer time compared to conventional pesticides which washes away in the rain. Ultimately, nanopesticides provide control on frequent use of chemical pesticides (Swamy and Asokan, 2013).

Conclusions

Bt based biopesticidal formulations will find wider application in future by adopting simple harvesting methods and robust and economical choice of various additives for different formulations. In general, there are two types of formulations, solid and liquid, including advances like encapsulations and both have a distinct application and hence equal markets. Various environmental factors namely, UV radiation, temperature, wind, pH and rain influence the field efficiency of Bt formulations. From late 1990s to recently, Bt formulation trends have progressed in the direction of "maximum efficacy per drop" resulting in high potency concentrates requiring lower spray volumes.

Reference

• Guerra, P.G., Michael, R., McGuire, Behle, R.W., Shasha and Wong, J.L., 2000, Microencapsulated formulations for improved residual activity of *Bacillus thuringiensis*. J. Econ. Entomol., **93** (2):219-225.

• Gunasekaran, G., Prabakaran, K. and Balaraman, 2002, Efficacy of a floating sustained release formulation of *Bacillus thuringiensis subsp. israelensis* in controlling *Culex quinqefaciatus* larvae in polluted water bodies. *Acta. Trop.,* **83**: 241-247.

• Navon, A., Keren, S., Levski, S., Grinstein, A. and Riven, Y., 1997. Granular feeding baits based on *Bacillus thuringiensis* products for the control of lepidopterous pests. *Phytoparasitica*, **25**: 1015-1105.

• Rowe, G.E. and Margaritis, A., 2004, Bioprocess design and economic analysis for the commercial production of environmentally friendly bioinsecticides from *Bacillus thuringiensis* HD- 1 *Krustaki. Biotechnol. Bioeng.,* **86**: 377-388.

• Satinder, K., Brar, M., Verma, R.D., Tyagi, J.R. and Valero, 2006, Recent advances in downstream processing and formulations of *Bacillus thuringiensis* based biopesticides Process. *Biochemistry*, **41**: 323-342.

• Swamy, H.M.H. and Asokan, R., 2013, *Bacillus thuringiensis* as 'Nanoparticles'- a Perspective for Crop Protection. *Nanoscience & Nanotechnology-Asia*, **3**: 102-105.

• Tsuji, K., 1997, Recent trends in pesticide formulations Association of Formulation Chemists, *Association of Formulation Chemists*, pp. 53-83.

15

Role of *Actinomycetes* in Insect Pest and Plant Disease Management

Mahesh S. Yandigeri

ICAR-National Bureau of Agriculture Insect Resources, Bangalore

Introduction

The actinomycetes are gram positive bacteria which have a characteristically high G+C content in their DNA (>55%). The name "Actinomycete" derives from the Greek *aktis* (a ray beam) and *mykes* (fungus) and was given to these organisms from initial observations of their morphology. The actinomycetes were originally considered to be an intermediate group between bacteria and fungi but now recognized as prokaryotic. They are phenotypically diverse and are found in most natural environments.

Actinomycetes are cosmopolitan in nature and encompass >80 genera and provide 80% of bioactive compounds as well as 3,500 antibiotic secondary metabolites. They are also used for management of pests and pathogens, bioleaching of metals, increasing soil fertility, generating biofuels, monitoring pollutants and waste treatment. The necessary extreme adaptions of extremophilic organisms to all aspects suggest that a wide variety of biomolecules may find application in existing and future biotechnological processes. *Streptomyces* species form a dominant group among the actinomycetes and have varied industrial as well as agricultural applications. They are also used as potential biological control agents to suppress a number of diseases.

Among actinobacteria, *Streptomyces* is the largest genus and the type genus of the family *Streptomycetaceae* (Fig. 1). Over 500 species of *Streptomyces* bacteria have been described. *Streptomyces* are predominantly in soil and decaying vegetation, most streptomycetes produce spores and are noted for their distinct "earthy" odor which results from production of a volatile metabolite, geosmin. *Streptomycetes* produce over two-thirds of the clinically useful antibiotics of natural origin (e.g., neomycin, chloramphenicol). The now uncommonly used streptomycin takes its name directly from *Streptomyces*. *Streptomycetes* are infrequent pathogens, though infections in human such as mycetoma can be caused by *S. somaliensis* and *S. sudanensis* and in plants can be caused by *S. caviscabies* and *S. scabies*.

Fig. 1. Typical *Streptomyces* mycelia and spores

Insect pests and diseases are the major impediment in enhancing the production of agricultural crops. The inundate use of synthetic pesticides has led to the development of pesticide-resistant pathogens and insect pests, causing environmental pollution, negative effects on natural enemies, human health hazards, and pollution of underground water, thereby causing ecological imbalance. The use of biological control and microbes having antimicrobial properties has become one of the most attractive options for enhancing the sustainability of agricultural production due to their ecofriendliness, low production cost, and reduced use of nonrenewable resources. Among microbes, actinomycetes are the decent alternative for the management of insect pests and diseases. These represent a high proportion of the soil microbial biomass and have the capacity to produce wide variety of secondary metabolites. These are the most economically and biotechnologically valuable prokaryotes. Several strains of actinomycetes have been acknowledged as abundant producer of valuable bioactive metabolites as antibacterial, antifungal, antibiotic, antiparasitic, insecticide, and herbicide. However, only a few microbial compounds are applicable at the field level or presently commercialized. Most of the actinomycetes are capable of producing secondary metabolites like antibiotics and antifungal compounds, especially those belonging to the genus *Streptomyces* and appear to be superior candidates to find new approaches for crop protection.

Actinomycetes as Bio-inoculants

Actinomycetes are ubiquitous in nature. They are found in soils, compost, freshwater basin, foodstuffs, and the atmosphere. These organisms exist and grow most profusely in different depths of soil and compost and in temperate and tropical regions all over the world. Actinomycetes belong to the order Actinomycetales, a gram-positive bacteria illustrated by a high genomic G +C content (74 mol %). Actinomycete species are distinguished as saprophytic bacteria that decompose organic matter, particularly biopolymers such as lignocellulose, starch, and chitin in soil and water. Several actinomycetes have typical biological features such as a mycelia growth and sporulation. They also hold the ability to biosynthesize a wide variety of antimicrobial compounds as secondary metabolites including agro-active compounds. Since the discovery of antibiotics namely streptomycin, actinomycete has received valuable interest and has resulted in the detection of diverse novel bioactive compounds of marketable value. A large number of actinomycetes have been isolated, characterized, and screened for their ability to produce commercially important compounds from different terrestrials. Among all, *Streptomyces* spp. are well known as a major source of bioactive natural products, which are mostly used in agrochemicals and pharmaceuticals. Streptomyces produce about 75 % of commercially useful antibiotics. Moreover, numerous species of the genus Streptomyces have established consideration due to their capability to produce a variety of secondary metabolites and bioactive compounds, including antibiotics and industrially important extracellular enzymes. Antifungal metabolite and extracellular hydrolytic enzyme production by different species of Streptomyces has been well explored by several researchers, under the major area of plant disease management. Many reports have illustrated the *in vitro* and *in vivo* antifungal potential of the actinomycetes (Table 1). Their modes of action include *via* enzymes such as cellulase, hemicellulase, chitinase, amylase, and glucanase, antagonism with pathogens, production of antibiotic, parasitism of hyphae, and siderophore production. A number of *Streptomyces* spp. are well known as antifungal biocontrol agents that inhibit numerous plant pathogenic fungi like *Phytophthora capsici, Fusarium oxysporum* f. sp. *cubense, Fusarium oxysporum* f. sp. *ciceri, Sclerotiumrolfsii, Alternaria alternata* and *Phomopsis archeri*, and *Rhizoctonia solani*.

All actinomycetes strain has possibly inherent potential for producing antimicrobial metabolites. Actinomycetes are used as plant growth-promoting agents, biocontrol tools, biopesticide agents, and antifungal compounds and as a source of agroactive compounds. Plant growth promotion potential of Streptomyces was reported on bean, tomato, wheat, and sorghum, rice, and chickpea. Actinomycetes produce many antibiotics including amphotericin, nystatin, chloramphenicol, gentamicin, erythromycin, vancomycin, tetracycline, novobiocin, and neomycin. Urauchimycins a member of antimycin class utilized as antifungal antibiotic against fungal pathogens and it act by hinders the electron flow in the mitochondrial respiratory chain.

Table 1. List of antagonistic actinomycetes and their disease-suppressing activity against plant pathogens

Actinomycetes	Plant	Disease	Pathogen
Streptoverticillium rimofaciens B-98891	Barley	Powdery mildew	*Erysiphe graminis* f. sp. *hordei*
Streptomyces viridodiasticus	Basal	Basal drop	*Sclerotinia minor*
Actinomadura roseola Ao108	Pepper	Blight	*Phytophthora capsici*
S. violaceusniger G10	Banana	Wilt	*Fusarium oxysporum* f. sp. *cubense* race 4
Streptomyces sp. KH-614	Rice	Blast	*Pyricularia oryzae*
Streptomyces sp. AP77	Porphyra	Red rot	*Pythium porphyrae*
Streptomyces sp. S30	Tomato	Damping-off	*Rhizoctonia solani*
S. halstedii	Red peppers	Blight	*P. capsici*
Streptomyces spp. 47W08, 47W10	Pepper	Blight	*P. capsici*
S. violaceusniger XL-2	-	Wood rot	*Phanerochaete chrysosporium, Postia placenta, Coriolus versicolor* and *Gloeophyllum trabeum*
S. ambofaciens S2	Red chilli fruits	Anthracnose	*C. gloeosporioides*
Streptomyces sp.	Sugar beet	Damping-off	*Sclerotium rolfsii*

S. hygroscopicus		Anthracnose and leaf blight	Colletotrichum gloeosporioides and S. rolfsii
Streptomyces sp.	Sunflower	Head and stem rot	Sclerotinia sclerotiorum
Streptomyces sp	Sweet pea	Powdery mildew	Oidium sp.
S. vinaceusdrappus	Rice	Blast	Curvularia oryzae, Pyricularia oryzae, Bipolaris oryzae and Fusarium oxysporum
Streptomyces sp. RO3	Lemon fruit	Green mold and sour rot	Penicillium digitatum and Geotrichum candidum
S. spororaveus RDS28	–	Root rot, collar or root rot, stalk rot, leaf spots, and gray mold rot or botrytis blight	R. solani, Fusarium solani, Fusarium verticillioides, Alternaria alternata, and Botrytis cinerea
S. toxytricini vh6	Tomato	Root rot	R. solani
Streptomyces spp.	Sugar beet	Root rot	R. solani and Phytophthora drechsleri
Streptomyces spp.	Chilli	Root rot, blight, and fruit rot	Alternaria brassiceae, Colletotrichum gloeosporioides, R. solani, and Phytophthora capsici
Streptomyces spp.	Chilli	Wilt	F. oxysporum f. sp. capsici

Source: Solanki et al. (2016)

Agro-active Metabolites and Antibiosis

Microorganisms serve as a pool of agro-active metabolites, as evident from the times. It has been estimated that approximately two-thirds of the thousands of naturally occurring antibiotics have been recovered from actinomycetes. It is known that, the proportion of all the actinomycetes that can be isolated from soil and other niches have the ability of producing antibiosis compounds such as volatiles, toxins, and antibiotics. It is a mechanism of biological control of plant disease that has been assessed in several systems. Researchers have carried systematic screening of antagonistic actinomycetes from soil for the production of antibiotics. An actinomycete from marine sediments of Andaman Islands has been identified with strong inhibitory activity against bacteria *Streptococcus, Staphylococcus aureus, Bacillus subtilis, Escherichia coli,* and *Proteus vulgaris* and fungi *Aspergillus niger, Candida albicans, Penicillium, Mucor,* and *Rhizopus* (Poosarla *et al.* 2013). Similarly, many actinomycetes have been found to be effective against a wide range of bacterial strains. *Streptomyces padanus*, recovered from the soil collected in Jiangxi Province, China, produced actinomycin X2, fungichromin, and a new polyene macrolide antibiotic which showed good antifungal activity. Antimycins have been identified from *Streptomyces* isolated from the integument of attine ants. *Streptomyces olivaceiscleroticus* AZ-SH514 and *Streptomyces antibioticus* AZ-Z710 produced antifungal compounds 4' phenyl- 1-napthyl-phenyl acetamide and mycangimycin.

Mycoparasitism/Hydrolytic Enzymes

The enzymes role in biological control is often deliberated by different mechanisms, parasitism and antibiosis in particular. Cell wall-degrading enzymes such as chitinase, β-1,3-glucanase, protease, and cellulase are important for mycoparasitism and antifungal activities. Actinomycetes are known to produce chitinase, β-1,3-glucanase, pectinase, xylanase, cellulase, amylase, protease, and lipase. Actinomycetes originated from agricultural soil have been producers of proteases, amylases, CMCase, xylanase, pectinase, and chitinase activities. Ten actinobacteria isolated from sediment samples of Kodiyakarai coast, the Bay of Bengal, India, exhibited multiple enzyme activity including amylase, cellulase, and protease (Manivasagan *et al.* 2010). Chitinase and glucanase are considered to be important hydrolytic enzymes in the lysis of fungal cell walls, for example, cell walls of *Fusarium oxysporum, Sclerotinia minor, S. rolfsii,* and *Aspergillus* (Singh *et al.* 1999; El-Tarabily *et al.* 2000; Hassan *et al.* 2011). Thirteen actinomycete strains were found to produce β-1,3-, β-1,4-, and β-1,6-glucanases, and these enzymes hydrolyze glucans from Phytophthora cell walls and cause lysis (Valois *et al.* 1996). Pattanapipitpaisal and Kamlandharn (2012) isolated 283 different chitinolytic actinomycete strains from rhizosphere-associated soils, from Ubon Ratchathani and Sisaket Province of Thailand, out of which 13 isolates have remarkably inhibited the growth of the fungus. Chitinases are group of the hydrolytic enzymes that catalyze depolymerization of chitin. After cellulose, chitin is the second most abundant organic compound in nature and is found to be rich in fungal cell walls. Among actinomycetes, species of the genus Streptomyces are well known for the production of chitinase, and hence the potential application of chitinase for biocontrol of fungal phytopathogens is promising (Gomes *et al.* 2000; Kim *et al.* 2003; Mukherjee and Sen 2006). The chitinase-producing strains could be used directly in biocontrol or indirectly by using purified proteins or through gene manipulation (Doumbou *et al.* 2001; Manivasagan *et al.* 2010; Sonia *et al.* 2011).

Root Colonizer and Plant Defense Activation

Roots operate a multitude of functions in plants including anchorage, nutrient and water acquisition, and production of exudates for plant development. The root–soil interface, or rhizosphere, is the reservoir of all the biological and chemical reactions within the soil matrix. Rhizosphere contains all kinds of microbes (beneficial and deleterious) with complex interactions. Deleterious microbes compete for nutrients with plant in rhizosphere and cause diseases, while PGPR support their host by nutrient mobilization and growth stimulation and protect the plant from biotic and abiotic stresses. PGPR are well known to regulate the plant health by controlling plant pathogens or *via* direct enhancement of plant development by providing nutrient. Literatures indicate that actinomycetes are playing an important role in the rhizosphere, where they may influence plant growth and protect plant roots against pathogen invasion by root. Root colonization is an essential character for the biocontrol agents against the pathogens, and higher colonization of biocontrol agents should reduce disease incidence. *Streptomyces* spp. 47W08 and 47W10 were used as protective agents against *Phytophthora capsici* in capsicum (Liang *et al.* 2005). Biocontrol bacteria have activated the plant defense system by producing peroxidase (POD), polyphenol oxidase (PPO), phenylalanine ammonia lyase (PAL), and superoxide dismutase (SOD) against pathogen invasion (Kim *et al.* 2007). POD, PPO, PAL, and SOD are strongly associated with plant disease resistance. POD catalyzes the lignin formation, enhances the thickness of plant cell wall to prevent pathogen invasion, and also balances the active oxygen metabolism. PPO oxidizes the phenols to quinone materials that have inhibitory effect on pathogen and also is involved in lignin synthesis. PAL is a rate-limiting enzyme that contributes in synthesis of phytoalexin, lignans, and phenolic compounds and promotes plant systemic resistance. SOD is an endogenous active oxygen scavenger in plants coupled with lignin synthesis. Host–microbe interaction is a very complex system. Lehr *et al.* (2008) have reported a complex interaction of *Streptomyces* spp. GB 4-2 with Norway spruce and *Heterobasidion abietinum*. GB 4-2 has promoted not only phytopathogenic fungus growth but also induced plant defense responses.

Host responses indicate that GB 4-2 induced both local and systemic defense responses in Norway spruce. *Streptomyces griseoviridis* is a superior model for colonization of plant rhizosphere by actinomycetes. *S. griseoviridis* is an antagonistic bacterium which has been isolated from light-colored Sphagnum peat (Tahvonen 1982) and is successful as biocontrol agent against the diseases such as damping-off of brassicas, fusarium wilt of carnation, and root rot of cucumber (Tahvonen and Lahdenpera, 1988). Kortemaa *et al.* (1994) have reported active root colonization of *S. griseoviridis* on turnip rape and carrot with higher colonization frequencies in turnip rape than carrot root. It concludes that colonization frequencies depend on host and environment factors. Different plant species produces various types and quantities of root exudates and metabolites, which positively affect the root colonization. The efficiency of *S. griseoviridis* bio-inoculum for seed dressing of barley and spring wheat against foot rot disease was investigated by Tahvonen *et al.* (1994) who observed higher yields in wheat than barley. Similarly, Cheng *et al.* (2014) observed colonization by *Streptomyces felleus* YJ1 against *Sclerotinia sclerotiorum* in oilseed rape and SOD, POD, PPO, and PAL activities.

Application of Actinomycetes Bio-inoculants

Against Fungal Plant Pathogens

Fungal plant pathogen causes serious damage in quantity and quality food production. The plant pathogens are controlled by chemical treatment; however, these chemicals also pose a negative impact on the environment and human health. Hence, microbe-based technology gained the attention to reduce the use of chemicals as they serve for both biocontrol and plant growth promotion. Golinska *et al.* (2015) have reported endophytic *Streptomyces* in enhancing the plant growth by nutrient mobilization and secondary metabolite production. Isono *et al.* (1965) discovered polyoxins A and B as new antifungal antibiotics from *Streptomyces cacaoi* var. *asoensis.* Iwasa *et al.* (1978) reported mildiomycin, a new antifungal compound isolated from *Streptoverticillium rimofaciens* B-98891, active against powdery mildew of barley. Chandra (1979) studied the mode of antifungal action of tetraene derived from *Streptomyces* sp. Rothrock and Gottlieb (1984) evaluated biocontrol activity of geldanamycin, a new antifungal agent from *S. hygroscopicus* var. *geldanus* and *S. griseus*, against Rhizoctonia root rot of pea. Tanaka *et al.* (1987) assessed globopeptin, a new antifungal antibiotic, and its *in vitro* antifungal activities against fungal pathogens. Novel antifungal antibiotics, phosmidosine, and their structure were studied and reported by Philips and McCloskey (1990). Matsuyama (1991) reported AC-1, an antifungal compound from *Streptomyces* sp. AB-88. Mand Nair *et al.* (1994) identified biocontrol application of gopalamicin against wheat powdery mildew, grape downy mildew, and rice blast pathogens. Tubercidin, a new antifungal compound reported by Kook and Kim (1995), was very effective against *Phytophthora capsici* blight in *Capsicum annuum*. Marten *et al.* (2001) reported the fungicidal activity of RhizovitR isolated from *Streptomyces rimosus* against *Pythium* spp., *Phytophthora* spp., *Rhizoctonia solani, Alternaria brassicicola*, and *Botrytis* sp. *Streptomyces violaceusniger* strain YCED-9, an antifungal biocontrol agent, produces three antimicrobial compounds (guanidyl fungin A, nigericin, and geldanamycin) against *Pythium* and *Phytophthora* spp. (Trejo-Estrada *et al.* 1998). Oligomycins A and C are macrolide antibiotics produced by *Streptomyces diastaticus* and 5 Actinomycetes Bio-inoculants: A Modern Prospectus for Plant Disease Management 69 exhibit a strong activity against *Aspergillus niger, A. alternata, Botrytis cinerea*, and *Phytophthora capsici* (Yang *et al.* 2010). Brief description of bioactive compounds produced by actinomycetes on various plant pathogens is given in Table 2. Besides production of antibiotic molecules, commercial bio-inoculants containing actinomycetes as active ingredients are also utilized for plant disease management. Cells of *Streptomyces griseoviridis* (Mycostop®) are used for the control of fusarium wilt of carnation and root rot disease of cucumber, and it has been used in greenhouse production to protect flowers from pathogens (White *et al.* 1990). Actinovate®, a biocontrol formulation of *S. lydicus* registered from AgBio in the United States of America, has been suggested for a wide range of environments ranging from greenhouses to field conditions. *S. lydicus* WYEC 108 (MicroPlus®) has been reported to possess disease suppression against powdery mildew and several root decay fungi.

Against Bacterial Plant Pathogens

Actinomycetes produce a broad spectrum of antimicrobial compounds, and these compounds are also useful for controlling bacterial diseases in different plants. Baz *et al.* (2012) reported 65–94 % reduction in the symptoms of disease severity caused by *Pectobacterium carotovorum* and *Pectobacterium atrosepticum*, causal agents of potato soft rot, by *Streptomyces* sp. strain OE7. Abdallah *et al.* (2013) studied biocontrol activity of actinomycete strains *Burkholderia cepacia* and *S. coelicolor* HHFA2 from Egyptian soils against onion bacterial rot diseases caused by *Erwinia carotovora* subsp. *carotovora* and observed significant reduction of disease incidence and enhancement of photosynthetic pigments. Hwang *et al.* (2001) explored the antimicrobial activity of phenylacetic acid and sodium phenylacetate isolated from *Streptomyces humidus* against the fungal and bacterial pathogens; both metabolites show inhibitory effect against *Saccharomyces cerevisiae* and *Pseudomonas syringae*pv. *syringae*. Lee *et al.* (2005) also reported multiple antimicrobial activity of 4-phenyl-3-butenoic acid against pathogenic fungus and bacteria in *in vitro* testing. *Streptomyces* sp. strain JJ45 showed antibiotic activity against the plant pathogenic bacteria *Xanthomonas campestris* pv. *campestris* and inhibitory compound identified as alpha-lsorbofuranose (3-->2)-beta-D-altrofuranose (Kang *et al.* 2009). Donghua *et al.* (2013) identified an antibacterial metabolite aloesaponarin II isolated from Streptomyces termitum ATC-2 that possessed strong antimicrobial activity against *Xanthomonas oryzae* pv. *oryzae* which causes bacterial blight in rice. Muangham *et al.* (2014) assessed a melanogenic actinomycete *Streptomyces bungoensis* TY68-3 for its ability to restrain the growth of *Xanthomonas oryzae* pv. *oryzae* and *Xanthomonas oryzae* pv. *oryzicola*. Mingma *et al.* (2014) reported inhibitory effect of *Streptomyces* sp. strain RM 365 against soybean pathogen *Xanthomonas campestris* pv. *glycines*.

Table 2. List of antifungal metabolites of actinomycetes assessed against different pathogens under *in vitro* and *in vivo* conditions

Metabolite	Source organism	Pathogen/disease
Polyoxins A and B	*Streptomyces cacaoi* var. *asoensis*	*Alternaria kikuchiana, Cochliobolus miyabeans, Pellicularia filamentosa* f. *sasakii, Pyricularia oryzae*
Mildiomycin	*Streptoverticillium rimofaciens* B-98891	*Rhodotorula rubra*
Tetraene	*Streptomyces* sp. A-7	*Helminthosporium oryzae, Curvularia lunata*
Geldanamycin	*S. hygroscopicus* var. *geldanus, S. griseus*	*R. solani*
Globopeptin	*Streptomyces* sp. MA-23	*Mucor racemosus, Pyricularia oryzae, B. cinerea,* and *A. kikuchiana*
Phosmidosine	*Streptomyces* sp. RK-16	*B. cinerea*
AC-1	*Streptomyces* sp. AB-88 M	*P. oryzae, B. cinerea, Helminthosporium maydis, H. oryzae,* and *Fusarium roseum* f. sp. *cerealis*
Gopalamicin	*Actinomycetes* MSU-625 and MSU-616	Wheat powdery mildew, grape downy mildew, and rice blast pathogens
Tubercidin	*Streptomyces violaceusniger*	*P. capsici, Magnaporthe grisea,* and *Colletotrichum gloeosporioides*

Manumycin	*Streptomyces flaveus* strain A-11	*P. capsici, M. grisea, Cladosporium cucumerinum,* and *Alternaria mali*
Streptimidone	*Micromonospora coerulea* strain Ao58	*P. capsici, M. grisea,* and *B. cinerea*
Daunomycin	*Actinomadura roseola* Ao108	*P. capsici* and *R. solani, Phytophthora*
Bafilomycins B1 and C1	*S. halstedii* K122	*Aspergillus fumigatus, Mucor hiemalis, Penicillium roqueforti,* and *Paecilomyces variotii*
Phenylacetic acid and sodium phenylacetate	*Streptomyces humidus* S5-55	*Pythium ultimum, P. capsici, R. solani, Saccharomyces cerevisiae,* and *Pseudomonas syringae* pv. *Syringae*
RhizovitR	*Streptomyces rimosus*	*Pythium* spp., *Phytophthora* spp., *R. solani, Alternaria brassicicola,* and *Botrytis* sp.
Fungichromin	*Streptomyces padanus* strain PMS-702	*R. solani*
4-Phenyl-3-butenoic acid	*Streptomyces koyangensis* strain VK-A60	*Colletotrichum orbiculare, M. grisea, Pythium ultimum, Pectobacterium carotovorum* subsp. *carotovorum,* and *Ralstonia solanacearum*
Antimycin A17	*Streptomyces* sp. GAAS7310	*Curvularia lunata, Rhizopus nigricans,* and *Colletotrichum nigrum*
Neopeptins	*Streptomyces* sp. KNF2047	*A. mali, B. cinerea, C. cucumerinum, Colletotrichum lagenarium, Didymella bryoniae,* and *M. grisea*
Natamycin	*Streptomyces lydicus* strain A01	*F. oxysporum, B. cinerea, Monilinia laxa*
5-Hydroxyl-5-methyl-2-hexenoic acid	*Actinoplanes* sp. HBDN08	*B. cinerea, C. cucumerinum,* and *Corynespora cassiicola*
Oligomycins A and C	*Streptomyces diastaticus*	*Aspergillus niger, Alternaria alternata, B. cinerea,* and *P. capsici*
Strevertenes	*Streptomyces psammoticus* KP1404	*A. mali, Aspergillus oryzae, Cylindrocarpon destructans, Colletotrichum orbiculare, F. oxysporum* f. sp. *lycopersici,* and *S. sclerotiorum*

Filipin III	*Streptomyces miharaensis* KPE62302H	*A. mali, A. niger, C. gloeosporioides, C. orbiculare, C. destructans, Diaporthe citri,* and *F. oxysporum*
Resistomycin and tetracenomycin D	*Streptomyces canus* BYB02	*M. grisea*
Antifungalmycin 702	*Streptomyces padanus* JAU4234	*M. grisea*
1H-Pyrrole-2-carboxylic acid (PCA)	*Streptomyces griseus* H7602	*P. capsici*
Bafilomycins B1 and C1	*Streptomyces cavourensis* NA4	*Fusarium* spp., *R. solani,* and *B. cinerea*

Source: Solanki *et al.* (2016)

Against Insect Pests

Many actinomycetes are utilized as pesticide, and at present, microbial insecticides are the main components of the biopesticide industry (Xiong *et al.* 2004). Biopesticides fall into three main groups: plant-incorporated protectants (PIPs), biochemical pesticides, and microbial pesticides. A microbial pesticide contains microorganisms as the active ingredient. Although each microbial active ingredient is comparatively specific for its target pest, it can also control various pests (Mazid *et al.* 2011; Ratna Kumari *et al.* 2014). Many reports are available for actinomycete's insecticidal activity including the boll weevil (Purcell *et al.* 1993), cotton leafworm, *Spodoptera littoralis* (Bream *et al.* 2001), *Culex quinquefasciatus* (Sundarapandian *et al.* 2002), housefly, *Musca domestica* (Hussain *et al.* 2002), *Drosophila melanogaster* (Gadelhak *et al.* 2005), *Helicoverpa armigera* (Xiong *et al.* 2004; Osman *et al.* 2007; Vijayabharathi *et al.* 2014), larvae of *Aedes aegypti* (Kekuda *et al.* 2010), Anopheles mosquito larvae (Dhanasekaran *et al.* 2010), and *Culex pipiens* (El–Khawagh *et al.* 2011). Bream *et al.* (2001) observed mortality of secondary metabolites of some actinomycete isolates including *Streptomyces* and *Streptoverticillium* on last instar larvae and pupae of the cotton leafworm *Spodoptera littoralis*. Osman *et al.* (2007) observed that cells of *Streptomyces* isolates were more active against cotton leafworm than culture filtrate. It shows that insecticidal activity present in both the actinomycete cell and cell filtrate could be utilized against insect pest. A macrotetrolide antibiotic identified from the acetone extract of *Streptomyces aureus* exhibited significant insecticidal activity against *Callosobruchus chinensis* (Oishi *et al.* 1970). The active compound quinomycin A extracted from ethyl acetate extract of *Streptomyces* sp. KN-0647 exhibited significant growth inhibition on the pathogenic insects *Aphis glycines, Culex pipiens, Dendrolimus punctatus, Plutella xylostella,* and *Spodoptera exigua* (Liu *et al.* 2008). Xiong *et al.* (2004) identified a strong insecticidal activity against both brine shrimp and *H. armigera* by avermectin B1 extracted from *Streptomyces* sp. 173. A new member of the tartrolone series of macrodiolides, tartrolone C, an insecticidal compound, was isolated from a *Streptomyces* sp. (Lewer *et al.* 2003). Kekuda *et al.* (2010) evaluated the larvicidal effect of two *Streptomyces* species isolated from the soil of Agumbe, Karnataka, India, against second instar larvae of *Aedes aegypti.* Selvakumar *et al.* (2011) reported entomopathogenic properties of *Brevibacterium frigoritolerans* against *Anomala dimidiata* and *Holotrichia longipennis,* and grub mortality occurred between the second and fifth weeks after inoculation under in vitro conditions. Sathya *et al.* (2016) reported a compound, diketopiperazine, cyclo(Trp-Phe), from *Streptomyces griseoplanus* SAI-25 that showed insecticidal activity against cotton bollworm, *Helicoverpa armigera.* Brief description of inhibitory compound of insect is given in Table 3.

Table 3. List of actinomycetes active against the insect pests and their inhibitory compounds

Actinomycetes	Insect pests	Inhibitory compound
Streptomyces aureus	*Callosobruchus chinensis*	Macrotetrolide antibiotic
Streptomyces sp.	Boll weevil	Protein
Streptomyces and *Streptoverticillium* spp.	*Spodoptera littoralis*	Secondary metabolite
Streptomyces 98-1	*Culex quinquefasciatus*	Extracellular metabolites
Streptomyces sp. 173	Brine shrimp and *Helicoverpa armigera*	Crude extract
Streptomyces spp.	*S. littoralis*	Cell protein
Streptomyces sp. KN-0647	*Spodoptera exigua, Dendrolimus punctatus, Plutella xylostella, Aphis glycines* and *Culex pipiens*	Quinomycin A
Streptomyces sp.	*Aedes aegypti*	Butanol extract
S. microflavus neau3	Adult mites and *Caenorhabditis elegans*	Macrocyclic lactone (1), isolated from fermented broth

Brevibacterium frigoritolerans	Anomala dimidiata and Holotrichia longipennis	Bacterial cells
S. bikiniensis A11	S. littoralis	Aminoglycoside antibiotic
Streptomyces sp. LC50	A. aegypti and brine shrimp	Crude extract
Streptomyces sp.	Sitophilus oryzae	Crude extract
Streptomyces sp. AP-123	H. armigera and Spodoptera litura	Polyketide metabolite
S. hydrogenans DH16	S. litura	Secondary metabolites in the fermentation broth
Streptomyces sp. AIAH-10	S. oryzae	Ethyl acetate extracts
S. griseoplanus SAI-25, S. bacillaris CAI-155, S. albolongus BCA-698	H. armigera, S. litura, and Chilo partellus	Extracellular metabolites
S. griseolus	Fasciola gigantica	Proteases

Advantages and Disadvantages of Actinomycetes Bio-inoculants

Several reports have discussed regarding the advantages and disadvantages of PGPR as bio-inoculants (Saharan and Nehra 2011; Trabelsi and Mhamdi 2013), and actinomycetes bio-inoculants are also one of them. Actinomycetes bio-inoculants and metabolites are naturally occurring substances that control pathogen and pests by nontoxic mechanisms. The beneficial effects of actinomycetes and their metabolites have been well assessed in the past; therefore, in recent times the agro-active antibiotics of actinomycetes are taking commercial importance in the market. Some actinomycetes are pathogenic in nature, so that regulatory regimes of most countries have actinomycetes inoculants banned in past time. Recently, considering the potential of the actinomycetes and their frequency and dominance in the agro-environment, it would be judicious to promote actinomycetes inoculants, after inclusive biosafety evaluation. The actinomycetes bio-inoculants advantages and disadvantage are described below.

Advantages

» It is naturally less harmful and eco-friendly and affects only specific pathogen or, in some cases, a few target organisms.

» It decomposes quickly, thereby resulting in lower exposures and largely avoiding the pollution problems.

» It supports the colonization of mycorrhizae.

» It balances the soil nutrient cycle and contributes to the residual pool of organic N and P, reducing N leaching loss and P fixation, and also supplies micronutrients to the plant to improve the metabolic activities.

» It provides food and supports the growth of beneficial insect, pest, and earthworms.

» They augment the plant defense and vice versa soil immunity to restrain the unwanted plant diseases, soil-borne diseases, and parasites.

» They normalize the plant metabolism against the biotic and abiotic stresses.

Disadvantages

» Proliferation rate is slow than other bacterial inoculants.

» Preparation and application is moderately different and susceptible to environmental factors.

» Success rate is not identical like chemical fertilizer.

» For storage, lower temperature is needed for longtime use.

References

• Abdallah, M.E., Haroun, S.A., Gomah, A.A., El–Naggar, N.E. and Badr, H.H., 2013. Application of actinomycetes as biocontrol agents in the management of onion bacterial rot diseases. *Arch. Phytopathol. Plant Protect.* **46** (15):1797–1808.

• Baz, M., Lahbabi, D., Samri, S., Val, F., Hamelin, G., Madore, I., Bouarab, K., Beaulieu, C. and Ennajiand, M.M., 2012. Control of potato soft rot caused by *Pectobacterium carotovorum* and *Pectobacterium atrosepticum* by Moroccan actinobacteria isolates. *World J. Microbiol. Biotechnol.*,**28**:303–311.

• Bream, A.S., Ghazal, S.A., Abd el–Aziz, Z.K. and Ibrahim, S.Y., 2001. Insecticidal activity of selected actinomycete strains against the Egyptian cotton leaf worm *Spodoptera littoralis* (Lepidoptera: Noctuidae). *Meded Rijksuniv Gent Fak Landbouwkd Toegep Biol. Wet.*,**66**:503–512.

• Chandra, A.L., 1979. Anti fungal activity of A-7, a new tetraene antibiotic. *Indian J. Exp. Biol.*, **3**:13–315.

- Cheng, G., Huang, Y., Yang, Y. and Liu, F., 2014. *Streptomyces felleus* YJ1: potential biocontrol agents against the sclerotinia stem rot (*Sclerotinia sclerotiorum*) of oilseed rape. *J. Agric. Sci.,* **6**:91–98.

- Dhanasekaran, D., Sakthi, V., Thajuddin, N. and Panneerselvam, A., 2010. Preliminary evaluation of anopheles mosquito larvicidal efficacy of mangrove actinobacteria. *Int J Appl. Biol. Pharm. Technol.,* **1**:374–381.

- Donghua, J., Qinying, L., Yiming, S. and Hao, J., 2013. Antimicrobial compound from a novel *Streptomyces termitum* strain ATC–2 against *Xanthomonas oryzae pv. oryzae. Res. J. Biotechnol.,* **8**:66–70.

- Doumbou, C.L., Hamby Salove, M.K., Crawford, D.L. and Beaulieu, C., 2001. Actinomycetes, promising tools to control plant diseases and to promote plant growth. *Phytoprotection,* **82**:85–102.

- El–Khawagh, M.A., Hamadah, K.S. and El–Sheikh, T.M., 2011. The insecticidal activity of actinomycetes metabolites, against the mosquito *Culex pipiens. Egypt Acad. J. Biolog. Sci.,* **4**:103–113.

- El–Tarabily, K.A., 2008. Promotion of tomato (*Lycopersicon esculentum* Mill.) plant growth by rhizosphere competent 1-aminocyclopropane-1-carboxylic acid deaminase producing streptomycete actinomycetes. *Plant Soil,* **308**:161–174.

- Gadelhak, G.G., EL–Tarabily, K.A. and AL–Kaabi, F.K., 2005. Insect control using chitinolytic soil actinomycetes as biocontrol agents. *Int. J. Agric. Biol.,* **7**:627–633.

- Golinska, P., Wypij, M., Agarkar, G., Rathod, D., Dahm, H. and Rai, M., 2015. *Endophytic actinobacteria* of medicinal plants: diversity and bioactivity. *Antonie Van Leeuwenhoek,* **108**:267–289.

- Gomes, R.C., Seme^do, L.T.A.S., Soares, R.M.A., Alviano, C.S., Linhares, L.F. and Coelho, R.R.R., 2000. Chitinolytic activity of actinomycetes from a cerrado soil and their potential in biocontrol. *Lett. Appl. Microbiol.* **30**:146–150.

- Hassan, A.A., El–Barawy, A.M. and Nahed, E.M.M., 2011. Evaluation of biological compounds of Streptomyces species for control of some fungal diseases. *J. Am. Sci.* **7**:752–760.

- Hussain, A.A., Mostafa, S.A., Ghazal, S.A. and Ibrahim, S.Y., 2002. Studies on antifungal antibiotic and bioinsecticidal activities of some actinomycete isolates. *Afr. J. Mycol. Biotechnol.*, **10**:63–80.

- Hwang, B.K., Lim, S.W., Kim, B.S., Lee, J.Y. and Moon, S.S., 2001. Isolation and in vivo and in vitro antifungal activity of phenylacetic acid and sodium phenylacetate from *Streptomyces humidus.Appl. Environ. Microbiol.*,**67**:3739–3745.

- Isono, K., Nagatsu, J., Kawashima, Y. and Suzuki, S., 1965. Studies on polyoxins, antifungal antibiotics. Part I. Isolation and characterization of polyoxins A and B. *Agric. Biol. Chem.*, **29**:848–854.

- Iwasa, T., Suetomi, K. and Kusuka, T., 1978. Taxonomic study and fermentation of producing organism and antimicrobial activity of mildiomycin.*J.Antibiot.*,**31**:511–518.

- Kang, Y.S., Lee, Y., Cho, S.K., Lee, K.H., Kim, B.J., Kim, M., Lim, Y. and Cho, M., 2009. Antibacterial activity of a disaccharide isolated from *Streptomyces* sp. strain JJ45 against *Xanthomonas* sp. *FEMS Microbiol. Lett.*,**294**:119–125.

- Kekuda, T.R.P., Shobha, K.S. and Onkarappa, R., 2010. Potent insecticidal activity of two *Streptomyces* species isolated from the soils of the Western Ghats of Agumbe, Karnataka.*J. Nat. Pharm.*,**1**:30–32.

- Kekuda, T.R.P., Shobha, K.S. and Onkarappa, R., 2010. Potent insecticidal activity of two *Streptomyces* species isolated from the soils of the Western Ghats of Agumbe, Karnataka.*J. Nat. Pharm.*,**1**:30–32.

- Kim, K.J., Yang, Y.J. and Kim, J.G., 2003. Purification and characterization of chitinase from *Streptomyces* sp. M-20.*J. Biochem. Mol. Biol.*,**36**:185–189.

- Kook, H.B. and Kim, B.S., 1995. In vivo efficacy and in vitro activity of tubercidin, an antibiotic nucleoside, for control of *Phytophthora capsici* blight in *Capsicum annuum.Pest Sci.*,**44**:255–260.

- Kortemaa, H., Rita, H., Haahtela, K. and Smolander, A., 1994. Root-colonization ability of antagonistic *Streptomyces griseoviridis*. *Plant Soil*, **163**:77–83.

- Lehr, N.A., Schrey, S.D., Hampp, R. and Tarkka, M.T., 2008. Root inoculation with a forest soil streptomycete leads to locally and systemically increased resistance against phytopathogens in Norway spruce. *New Phytol.*,**177**:965–976.

- Lewer, P., Chapin, E.L., Graupner, P.R., Gilbert, J.R. and Peacock, C., 2003. Tartrolone C: a novel insecticidal macrodiolide produced by *Streptomyces* sp. CP1130. *J. Nat. Prod.*, **66**:143–145.

- Liu, H., Qin, S., Wang, Y., Li, W. and Zhang, J., 2008. Insecticidal action of quinomycin A from *Streptomyces* sp KN-0647 isolated from a forest soil. World. *J. Microbiol. Biotechnol.*,**24**:2243–2248.

- Manivasagan, P., Gnanam, S., Sivakumar, K., Thangaradjou, T., Vijayalakshmi, S. and Balasubramanian, T., 2010. Isolation, identification and characterization of multiple enzyme producing actinobacteria from sediment samples of Kodiyakarai coast, the Bay of Bengal. *Afr. J. Microbiol. Res.*,**4**:1550–1559.

- Marten, P., Bruckner, S., Minkwitz, A., Luth, P. and Bergm, G., 2001. RhizovitR: impact and formulation of a new bacterial product. In: Koch E, Leinonen P (eds) Formulation of microbial inoculants: proceedings of a meeting held in Braunschweig, Germany. COST Action 830/Microbial inoculants for agriculture and environment, Germany, pp 78–82.

- Matsuyama, N., 1991. Purification and characterization of antifungal substance AC-1 produced by a *Streptomyces* sp. AB–88 M. *Ann. Phytophathol. Soc. Jpn.*,**57**:591–594.

- Mazid, S., Kalita, J.C. and Rajkhowa, R.C., 2011. A review on the use of biopesticides in insect pest management. *Int. J. Sci. Adv. Technol.*,**1**(7):169–178.

- Mingma, R., Pathom–aree, W., Trakulnaleamsai, S., Thamchaipenet, A. and Duangmal, K., 2014. Isolation of rhizospheric and roots endophytic actinomycetes from leguminosae plant and their activities to inhibit soybean pathogen, *Xanthomonas campestris*pv. *glycine*. *World J. Microbiol. Biotechnol.*,**30**:271–280.

- Muangham, S., Pathom–aree, W. and Duangmal, K., 2014. Melanogenic actinomycetes from rhizosphere soil -antagonistic activity against *Xanthomonas oryzae* and plant-growth-promoting traits. *Can. J. Microbiol.*,**61**:164–170.

- Mukherjee, G. and Sen, S.K., 2006. Purification, characterization and antifungal activity of chitinase from *Streptomyces venezuelae* P10. *Curr. Microbiol.*,**53**:265–269.

- Nair, M.G., Amitabh, C., Thorogod, D.L. and Chandra, A., 1994. Gopalamicin, an antifungal macrodiolide produced by soil actinomycetes. *J. Agric. Food Chem.*, **42**:2308–2310.

- Oishi, H., Sugawa, T., Okutomi, T., Suzuki, K., Hayashi, T., Sawada, M. and Ando, K., 1970. Insecticidal activity of macrotetrolide antibiotics. *J. Antibiot.,***23**:105–106.

- Osman, G., Mostafa, S. and Mohamed, S.H., 2007. Antagonistic and insecticidal activities of some *Streptomyces* isolates. *Pak. J. Biotechnol.*, **4**:65–71.

- Pattanapipitpaisal, P. and Kamlandharn, R., 2012. Screening of chitinolytic actinomycetes for biological control of *Sclerotium rolfsii* stem rot disease of chilli. *Songklanakarin. J. Sci. Technol.,***34**:387–393.

- Philips, D.R. and Mc Closkey, J.A., 1990. Isolation and characterization of phosmidosine. A new antifungal nucleotide antibiotic. *J. Antibiot.,***44**:375–381.

- Poosarla, A., Ramana, L.V. and Krishna, R.M., 2013. Isolation of potent antibiotic producing actinomycetes from marine sediments of Andaman and Nicobar Marine Islands. *J. Microbiol. Antimicrob.,***5**:6–12.

- Purcell, J.P., Greenplate, J.T., Jennings, M.G., Ryerse, J.S., Pershing, J.C., Sims, S.R., Prinsen, M.J., Corbin, D.R., Tran, M., Sammons, R.D. and Stonard, R.J., 1993. Cholesterol oxidase: a potent insecticidal protein active against boll weevil larvae. *Biochem. Biophys. Res. Commun.*, **196**:1406–1413.

- Ratna Kumari, B., Vijayabharathi, R., Srinivas, V. and Gopalakrishnan, S., 2014. Microbes as interesting source of novel insecticides: a review. *Afr. J. Biotechnol.***13**:2582–2592.

- Rothrock, C.S. and Gottlieb, D., 1984. Role of antibiosis in antagonism of *Streptomyces hygroscopicus* var. *geldanus* to *Rhizoctonia solani* in soil. *Can. J. Microbiol.,***30**:1440–1447.

- Saharan, B.S. and Nehra, V., 2011. Plant growth promoting rhizobacteria: a critical review. *Life Sci. Med. Res.,***21**:1–30.

- Sathya, A., Vijayabharathi, R., Vadlamudi, S., Sharma, H.C. and Gopalakrishnan, S., 2016. Assessment of tryptophan based diketopiperazine, cyclo (L-Trp-L-Phe) from *Streptomyces griseoplanus* SAI-25 against *Helicoverpa armigera* (Hu¨bner). *J. Appl. Entomol. Zool.*, **51**(01):01–10. ISSN 0003-6862.

- Selvakumar, G., Sushil, S.N., Stanley, J., Mohan, M., Deol, A. and Rai, D., 2011. *Brevibacterium frigoritolerans* a novel entomopathogen of *Anomala dimidiata* and *Holotrichia longipennis* (Scarabaeidae: Coleoptera). *Biocontrol. Sci. Technol.,***21**:821–827.

• Singh, P.P., Shin, Y.C., Park, C.S. and Chung, Y.R., 1999. Biological control of *Fusarium* wilt of cucumber by chitinolytic bacteria. *Phytopathology,* **89**:92–99.

• Solanki, M.K., Singh, R.K., Srivastava, S., Kumar, S., Srivastava, A.K., Kashyup, P.L. and Arora, D.K., 2013. Isolation and characterizations of siderophore producing rhizobacteria against *Rhizoctonia solani. J. Basic Microbiol.* **54**:585–597.

• Sonia, M.T., Jedidi, N. and Abdennaceur, H., 2011. Studies on the ecology of actinomycetes in an agricultural soil amended with organic residues: I. identification of the 80 M.K. Solanki et al. dominant groups of actinomycetales. *World J. Microbiol. Biotechnol.* **27**:2239–2249.

• Sundarapandian, S., Sundaram, M.D., Tholkappian, P. and Balasubramanian, V., 2002. Mosquitocidal properties of indigenous fungi and actinomycetes against *Culex quinquefasciatus. J. Biol. Control,* **16**:89–91.

• Tahvonen, R., Hannukkala, A. and Avikainen, H., 1994. Effect of seed dressing treatment of *Streptomyces griseoviridis* on barley and spring wheat in field experiments. *Agric. Sci. Fini.,* **4**:419–427.

• Tahvonen, R. and Lahdenpera, M.L., 1988. Biological control of *Botrytis cinerea* and *Rhizoctonia solani* in lettuce by *Streptomyces* sp. *Ann. Agric. Fenn.,* **27**:107–116.

• Tanaka, Y., Hirata, K., Takahashi, Y., Iwai, Y. and Omura, S., 1987. Globopeptin, a new antifungal peptide antibiotic. *J. Antibiot.,* **40**:242–244.

• Trejo-Estrada, S.R., Sepulveda, I.R. and Crawford, D.L., 1998. In vitro and in vivo antagonism of *Streptomyces violaceusniger* YCED9 against fungal pathogens of turfgrass. *World J. Microbiol. Biotechnol.,* **14**:865–872.

• Valois, D., Fayad, K., Barasubiye, T., Garon, M., Dery, C., Brzezinski, R. and Beaulieu, C., 1996. Glucanolytic actinomycetes antagonistic to *Phytophthora fragariae* var. rubi, the causal agent of raspberry root rot. *App. Environ. Microbiol.* **62**:1630–1635.

• Vijayabharathi, R., Kumari, B.R., Sathya, A., Srinivas, V., Abhishek, R., Sharma, H.C. and Gopalakrishnan, S. 2014. Biological activity of entomopathogenic actinomycetes against lepidopteran insects (Noctuidae: Lepidoptera). *Can. J. Plant Sci.,* **94**:759–769.

- White, J.G., Linfield, C.A., Lahdenpera, M.L. and Uoti, J., 1990. Mycostop, a novel biofungicide based on *Streptomyces griseoviridis*. In: Brighton crop protection council, pests and diseases, pp 221–226.

- Xiong, L., Li, J. and Kong, F., 2004. *Streptomyces* sp. 173, an insecticidal microorganism from marine. *Lett. Appl. Microbiol.* **38**:32–37.

- Yang, P.W., Li, M.G., Zhao, J.Y., Zhu, M.Z., Shang, H., Li, J.R., Cui, X.L., Huang, R. and Wen, M.L., 2010. Oligomycins A and C, major secondary metabolites isolated from the newly isolated strain *Streptomyces diastaticus*. *Folia Microbiol.* **55**:10–16.

16

Synergistic and Antagonistic Response of Bioagents with Agro Chemicals

**Kiran Kumar K.C.*, Raghavendra Achari,
Jayasudha S.M. and Goureesh Bhat**
University of Horticulture Sciences, Bagalkot

Diseases are one of the major constraints in crop production since the beginning of the agriculture. The innovations in the field of plant pathology has worked out lot of options for safe management of plant diseases. The use of bio agents is one of the economical and eco-friendly way of management of crop diseases. The sole use of bioagents will not be able to manage the disease effectively. Most of the time bioagents find a place under Integrated Disease Management strategy. Some time, they need to be alternated with pesticide molecules.

In recent years, farming sector is facing shortage of labour for farm operations due to migration of people to cities for better job opportunities and perks. To overcome the problem of labour shortage, farmers have started spraying the mixture of fungicides, insecticides, foliar nutrients and bio agents as a tank mix instead spraying them individually with an objective to reduce the number of sprays, time, energy, equipment cost and thereby can overcome the shortage of farm labour. Not all changes are for the better pesticides in combinations usually alter plant absorption and translocation as well as metabolism and sometimes causing toxicity at their site of action. Negative effects can occur such as reduced pest control, increased damage to non-target plants (phytotoxicity) and incompatibility problems among the chemical molecules mixed before used. Therefore, to use mixture of these agrochemicals and biological agents, it is essential to study their compatibility before taking up the spray on the target crop.

Although use of biocontrol agents could reduce chemical application to a limited extent, as it is less efficient (Monte, 2001).One of the most promising possibilities for the application of biocontrol agent *Trichoderma* strains is within the frames of a complex integrated plant protection modules, which are based on the combined application of physical, chemical and biological components of control. In case of the adoption of a complex integrated management strategy wherein *Trichoderma* strains are need to be combined with chemical pesticides, it is important to collect information about effect of pesticides on the biocontrol agent (Kredics *et al.*, 2003).The combined use of biocontrol agents and chemical pesticides has attracted much attention as a way to obtain synergistic or additive effects in the control of plant diseases, especially soil-borne pathogens. Though mixing of fungicides with insecticides was first practiced by Weed during 1889 itself, but work in this line was not gained importance in the past and hence information available on this aspect is very scanty. In view of scarcity of labours for farm operations in recent years, studies on the cocktail spray of pesticides, bio pesticides and foliar nutrients is taking due importance in farm sector.

Compatibility: Compatibility is a state in which two things are able to exist or occur together without problems or conflicts. Compatibility is the ability of two or more components of a pesticide mixture to be used in combinations without impairment of toxicity, physical properties or plant safety of either of the component.

Types of Compatibility

The interaction between the biocontrol agents and the agrochemicals may be explained in terms of 1. Additive effect 2.Synergism 3. Antagonism and 4. Enhancement

Pesticides to be used in combination or as components in a mixture may be either compatible or incompatible [Horsfall *et al.*, 1945]. Incompatible refers to the reaction of pesticides that cannot be mixed safely without impairing the effectiveness of either one or both chemicals, developing undesirable physical or chemical properties or causing plant injury (Phytotoxic), whereas in case of compatible chemicals, that can be mixed together without causing any problem on the sprayed plants without compromising the efficacy of all the ingredients mixed.

The ingredients of most of the pesticides are highly active substances and mixtures of such materials are likely to react and undergo physical and/or chemical changes. Physical incompatibility involves an unstable mixture that settles out by precipitation, flocculates, lumps, globs, sediments, gums, foams excessively or disperses poorly and reduces efficiency and clogging of sprayer nozzles. This type of incompatibility may also happen due to the use of hard/cold water for preparing the fungicide/foliar nutrients solution. Chemical incompatibility involves the breakdown and loss of effectiveness of one or more products in the spray tank and possible formation of one chemical that are insoluble or phytotoxic. However the incompatibleness with biological sources leads to biological incompatibility (Bhat *et al.*, 2000).

Benefits of mixing of biocontrol agents with pesticides

The combined use of biocontrol agent and chemical pesticides has attracted much attention in order to obtain synergistic or additive effects in the control of soil borne diseases (Locke *et al.*, 1985). Reduced amount of fungicides can stress and weaken the pathogen and render its propogules more susceptible to subsequent attack by the antagonist (Heiljord and Tronsmo, 1998). Srinivas and Ramkrishnan (2002) have reported that integration of biocontrol agents and commonly used fungicides showed positive association by reducing the seed infection compared to fungicide and the fungal antagonists' when used individually. Recently, biological control combined with chemical fungicide at lower concentration is applicable. It has become a part of IPM (Integrated Plant disease Management) strategy which was applied since sole use of chemicals known to cause environmental pollution.

How to evaluate the compatibility of bioagents...?

a. *In-vitro* condition

1. Compatibility of fungal biocontrol agents

Poisoned food technique

The fungal biological control agent has to be grown on PDA medium in Petriplates for 10 days prior to setting up of experiment. Suspension has to be prepared in PDA by adding required quantity of fungicide to obtain the desired concentration on the basis of active ingredient present in the formulation used. 20 ml of poisoned medium has to be poured in each of sterilized Petriplates. Suitable checks need to be maintained without adding fungicides. 5 mm mycelial disc of organism interested need to be taken from periphery of 10 days old colony and has to be placed on to solidified medium in the centre and incubate at 28±1⁰C for 15 days. Each treatment need to be replicated suitably to satisfy the requirement of statistical analysis. The diameter of radial growth of fungal mycelium has to be measured in two directions and the average should be taken. Percentage of inhibition of growth has to worked out by the formula given below, (Wincent, 1947 and Sharvelle, 1951).

$$I = \frac{C-T}{C} \times 100$$

Wherein, I is inhibition, C is radial growth of mycelia as diameter (mm) in control plates, whereas, T is radial growth of mycelia as diameter (mm) in treatment plates.

Compatibility of bacterial biocontrol agents

Turbidometric method:

Culture of bacterial biocontrol agents (1ml) has to be transferred to 100 ml flask containing 50 ml of King's B/ nutrient broth. The calculated quantity of chemical/ bacteriocides has to be amended in the broth. Need to keep a standard check and also negative control without adding any chemical. The flasks need to be incubated at 28±1⁰C in Psychotherm shaker and determine the optical density value of the culture broth using spectophotocolorimeter 610 nm at regularly at 6h intervals (Valarmathi *et al.*, 2013).

b. Pot and field condition:

Earthen pots filled with steam sterilized soil mixture need to be used for growing the plants. The rhizosphere soil has to be inoculated with virulent culture of the pathogen to make the plants infected. The interested biocontrol agent has to be incorporated to the rhizosphere either alone as well as in combination with agro chemicals needs to be evaluated as different treatments. Before this it is necessary to standardise the effective dose of the bio control agents and required quantity of chemical concentrations. The above treatments need to be repeated and plants need to be maintained with proper cultivation practices. Untreated plants act as control for both bio control agents as well as pathogens. Observations need to be recorded at frequent intervals for the growth characters and disease symptoms. In the similar pattern, the experiments under field condition also can be carried out with the natural infection of plants or in sick plots (Jayakumar and Ramakrishnan, 2012).

c. Compatibility of formulated products (Jar test)

1. Measure 500 ml of water into a one litre glass jar. This should be the same water you would fill a spray tank.

2. Add ingredients and stirring after each addition.

3. Let the solution stand in a ventilated area for 15 minutes and observe the results. If the mixture is giving off heat, these ingredients are not compatible. If gel or scum forms or solids settle at the bottom (except for the wettable powders) then the mixture is likely not compatible.

4. If no signs of physical incompatibility appear, test the mixture using a spray bottle on a small area where it is to be applied. Look for phytotoxic indications, such as plant damage, and monitor efficacy (which is hard to do unless you actually fill the sprayer and try it on a few plants).

Compatibility of biological agents with fungicides

Arunasri (2003) reported that the *in vitro* compatibility of four fungicides tested on *Trichoderma* isolate-1 (T1) and *Pseudomonas* sp. (B1) at five concentrations *viz.*, 50, 100, 250, 500 and 1000 ppm using poisoned food technique and inhibition zone technique. Thiram was found to be best compatible fungicide with both *Trichoderma* isolate-1 (T1) and *Pseudomonas* sp. (B1) as it inhibited growth to minimum extent of 31.39% of T1 and 2.22% of B1.

Durai (2004) evaluated two systemic and non-systemic fungicides for their compatibility with *P.fluorescens*. Of the fungicides tested, thiram at 1000 ppm shown the highest inhibition (4.7%). Mancozeb was found compatible with *P. fluorescens* (B1) even at 50 ppm concentration as the growth inhibition was nil, the inhibition was only 0.7 per cent even at 1000 ppm concentration.

a. Seed treatment

Prasanna *et al.* (2002) reported that reduction in the radial growth of *Rhizoctonia solani* was 40.2 per cent in presence of *T. harzianum*, whereas no inhibition was observed in thiomethoxam at 10 g kg-1 seed *in vitro*. In combination with thiomethoxam at 2.85 g kg-1 seed + *Trichoderma*, 4.28 g kg-1 seed + *Trichoderma* and 10 g kg-1 seed + *Trichoderma*, the reduction in radial growth of *R. solani* was 37.3, 35.95 and 32.4 per cent, respectively.

Durai (2004) conducted pot culture experiments on integrated disease management for charcoal rot of sesamum under glasshouse conditions. Out of 15 treatment combinations evaluated, maximum disease reduction was achieved by integrated use of seed treatment with mancozeb (100 ppm) + soil application of native *T.viride* (10 g/kg soil) and *Pseudomonas fluorescens* (20 ml/kg of soil). As this treatment recorded PDI of 10.67 only. This treatment not only reduced the disease incidence to a maximum extent, but also recorded maximum plant height (22.29 cm), shoot length (5.41 cm), and root length (4.27 cm), maximum dry weight of shoot (76.28 mg) and root (52.31 mg) of sesame when compared to all other treatments.

b. Soil borne disease

Rama Bhadra Raju and Raoof (2003) indicated that the improved strain of *T. viride* (B-16) tolerant to carbendazim developed at DOR, Hyderabad was evaluated for the management of *Fusarium oxysporum* f.sp. *ricini*, causing castor wilt. The effectiveness of B-16 increased with the quantity used as seed treatment. B-16 seed treatment at 5, 7.5 and 10 g kg-1 resulted in 18.7, 23.4 and 30.4 per cent reduction in wilt, incidence over control respectively. Seed dressing with Carbendazim (0.2%) and B-16 (10 g kg-1) combined soil application of B-16 @ 190 kg ha-1 (2.5 kg of formulation in 190 kg of farmyard manure) resulted in effective wilt control, where the proliferation of *T. viride* was maximum (18 x 103 cfu g-1) and significant increase in yield (293.8%) over control was recorded. The population build up of *T. viride* was least in Carbendazim treatment (3.8 x 103 cfu g-1).

c. Post harvest disease

Pavanello et al. (2015), evaluated the effect of the application of *T. harzianum* in combination with different fungicides applied before harvest to 'Eldorado' peaches for brown rot control and other quality parameters during storage. The treatments consisted of five preharvest fungicide applications (control, captan, iprodione, iminoctadine and tebuconazole) associated with postharvest application of *T. harzianum*, after cold storage (with and without application), in three evaluation times (zero, two and four days at 20 °C), The application of *T. harzianum* only brought benefits to the control of brown rot when combined with the fungicide captan, at zero day shelf life. After two days, there was a greater skin darkening in peaches treated with *T. harzianum* compared with peaches without the treatment, except for peaches treated with the fungicide iprodione and *T. harzianum*. The application of T. *harzianum* during postharvest stage showed no benefits for the control of brown rot, however, the along with fungicides reduced the incidence of *Rhizopus stolonifer* during the shelf life.

Compatibility of biological agents with bactericides

El-Khair *et al.*, (2007) evaluated the bactericides, i.e., streptomycin sulfate, Starner and Micronite Soreil and two bioagents, *Tricoderma harzianum* and *Bacillus subtilis* for controlling the soft rot disease causing by *Erwinia carotovora* subsp. *carotovora* under *in vitro* and in field. *In vitro*, results showed that the Starner, *B. subtilis* and *T. harzianum* reduced the pectolytic enzymes (PG and PME enzymes), while Starner and streptomycin sulfate reduced the cellulolytic enzyme (Cx). The tested materials were also powerful bactericide against the bacterial soft rot pathogen. Streptomycin sulfate, *T. harizanum* and *B. subtilis* prevented the soft rot disease in daughter potato tubers and increased the vegetative characters, like plant height and number of leaves per plant. Results show that plant tubers yield and the average tubers weight has been increased when the above bactericides were applied, comparing with un-treated plants. Starner and Micronite Soreil gave a moderate effect in reducing the incidence of soft rot disease, while a positive effect on tuber weight and plant tuber yield has been recorded than control. The incidence of soft rot disease and the weight loss in potato tubers resulting from treated plants were studied in storage.

Compatibility of biological agents with nematicide

Jayakumar and Ramakrishnan (2009) studied on compatibility of avermectin with carbofuran 3G, *Pseudomonas fluorescens* and *Trichoderma viride* for the management of *Meloidogyne incognita* in tomato and revealed that combination of seedling root dip with avermectin 75% and soil application of carbofuran 3G @ 1kg a.i ha-1 recorded maximum shoot length of 36.90cm, fresh shoot weight of 16.80 g, dry shoot weight of 5.2g, and fruit yield of 290.50g per plant. It was followed by avermectin + *P. fluorescens* and avermectin + *T. viride*. Maximum reduction in number of *M. incognita* adult females (14.33), number of egg masses per g root weight (6.33), number of eggs per egg mass (105.33), soil nematode population (105.67) and root knot index (1.0) was recorded in plants treated with avermectin + carbofuran 3G, followed by combined application of avermectin with *P. fluorescens* or *T. viride*.

Compatibility of biological agents with insecticide

Mukhopadyay *et al*. (1986) tested the effect of metalaxyl on radial growth of *T. harzianum* at 0.1, 1, 10 and 100 ppm concentrations and reported that there was no effect of metalaxyl on the growth and sporulation of *T. harzianum* upto 100 ppm concentration.

Compatibility of biological agents with herbicides

T. harzianum was able to grow in the presence of the herbicide, diallate (Domsch *et al*., 1980), Sharma and Mishra (1995) observed better tolerance of a strain of *T. harzianum* to 2, 4-D as compared to fluchloralin and pendimethalin. The tolerance of *Trichoderma* to certain herbicides might be attributed to gradual utilization of herbicides as a source of C or N by the organism. This has been observed in a strain of *Trichoderma viride* which grown on simazine and atrazine a sole C or N source (Domsch *et al*., 1980). Presence of proper enzyme system to disintegrate the toxic residue to basic non-toxic moieties.

Compatibility of biological agents with organic and in- organic fertilizers

Jayaraj and Ramabhadran (1998) tested the seven nitrogenous salts, ammonium nitrate, calcium nitrate, sodium nitrate, potassium nitrate, ammonium sulphate, ammonium chloride and urea under *in vitro* for their effect on growth, sporulation, production of cellulase and antifungal substances (AFS) by *Trichoderma harzianum*. It was reported that N sources at 0.2% level (w/v), ammonium nitrate recorded the maximum mycelial dry weight (500 mg), followed by sodium nitrate (450 mg) and ammonium sulphate (425 mg). Urea and calcium nitrate recorded the least growth, but maximum sporulation occurred in ammonium sulphate.

Shylajal and Rao (2012) tested the *Trichoderma harziaum* under *in- vitro* for its compatibility with different concentrations of commonly used inorganic fertilizers. Four different inorganic fertilizers *viz.*, urea, single super phosphate (SSP), muriate of potash (MoP) and calcium ammonium nitrate (CAN) were used, each at concentrations of 100, 200, 500, 1000 and 2000 ppm. Urea at 1000 ppm and above increased the colony diameter of *T. harziaum* by 11.1%. MoP increased the growth of the biological control agent at all concentrations tested while SSP and CAN both inhibited it. The inhibition ranged from 8.8% to 13% for SSP and from 11.1% to 71.9% for CAN and increased with the increase in concentration.

Naznin *et al.* (2015) conducted an experiment to determine the appropriate dose and combination of organic and chemical fertilizers and to assess the effect of bio-control agent (*Trichoderma*) on qualitative and quantitative characteristics of tuberose (*Polianthes tuberosa* L. cv. Single), including stem length, rachis length, spike length, floret number, flower yield, flower durability, number of bulb etc. The experiment was laid out in Randomized Complete Block Design (RCBD) with three replications having eight treatments as follows: T1= Farmyard manure (5 t/ha) + ¼ RDF, T2= Poultry refuse (5 t/ha) + ¼ RDF, T3= Bokashi (3 t/ha) + ¼ RDF, T4= Mustard oil cake (500 kg/ha) + ¼ RDF, T5= Vermicompost (5 t/ha) + ¼ RDF, T6= Tricho-compost (3 t/ha) + ¼ RDF, T7= Tricho-leachate (3000 L/ha) + ¼ RDF and T8= Control (Recommended doses of fertilizer) (N150 P45 K88 S10 B1 Zn1 kg/ha). Maximum growth, yield and yield contributing characters were recorded in T6= Tricho-compost (3 t/ha) + ¼ RDF which were statistically superior to other treatments. Maximum plants emergence (93.3%) recorded in T6 (Tricho-compost + ¼ RDF). In case of plant height, number of leaves per plant, plant spread, days to flowering, number of florets, flower yield, bulb production, T6= Tricho-compost (3 t/ha) + ¼ RDF gave superior results over control (Recommended doses of fertilizer). The data obtained from the experiment showed that Tricho-compost with fertilizer enhanced qualitative and quantitative characters of tuberose flowers.

Compatibility of among the bio-agnets

Prasanthi *et al.* (2000) evaluated certain fungal and bacterial biocontrol agents as seed and soil application against safflower root rot caused by *R. bataticola*. Both seed treatment and soil drenching with antagonists increased safflower seedling percentage survival, seed treatment being more effective than soil drenching with highest survival rate with *T. viride* (83.33%) and *Pseudomonas fluorescens* (86.66%).

Conclusion

In the view of the inherent hazardous effects involved in conventional chemical management and with the introduction of organic farming, biological control of plant diseases using antagonists is the distinct possibility. However, the interaction of these biocontrol agents with agrochemicals is essential since the present scenario of plant disease management is through IDM practices. Biological control agents are living organisms that can be inhibited or killed by pesticides. Many of these will fail if growers also make pesticide applications to the same set of plants before, during, or after application of biocontrol agents. It is very important to avoid pesticide use 3-4 months prior to using biological control agents. It is necessary to evaluate their compatibleness with the routinely using agrochemicals in that specified cropping system. Further, there is appreciable work need to be carried out towards identification or artificial improvement of pesticide tolerant strains.

References

- Anonymous 1988, Problems of mixing pesticides.In: *Report on Plant Disease* (RPD No.1004), University of Extension Agricultural Consumer and Environmental Sciences., 1-5.

- Arunasri, 2003. Management of collar rot disease of crossandra [*Crossandra infundibuliformis* (L.) Nees] incited by *Sclerotium rolfsii* Sacc. M Sc (Ag.) *Thesis* Acharya N G Ranga Agricultural University, Hyderabad.

- Bhat, S. S., Naidu, R. and Daivasikamani, S., 2000 Integrated disease management in coffee.In:*IPM system in Agriculture – Cash crops,* Upadhyay RK, Mukerji KG, Dubey OP, (ed.), Aditya Books Private Limited, New Delhi, India, , 65-82.

- Bicci, M., Cinar, O. and Erkilic, A., 1994. Cultural, chemical, physical and biological methods to control stem rot, *Sclerotium rolfsii* in peanut. *Turkish. Journal of Agriculture and Forestry,* **18** (5): 423-435.

- Domsch, K. H. 1980. Influence of selected pesticides on the microbial degradation of 14 C-triallate and 14 C-diallate in soil. Archives of Environmental Contamination and Toxicology, 9(1); 115-123.

- Durai, M., 2004. Management of charcoal rot of sesame (*Sesamum indicum* L.) incited by *Macrophomina phaseolina* (Tassi.) Goid.*M Sc (Ag.) thesis, Acharya N*

G Ranga Agri. Univ. Hyderabad.

- El-Khair, Abd, Karima, H. and Haggag, H. E., 2007. Application of Some Bactericides and Bioagents for Controlling the Soft Rot Disease in Potato. *Research Journal of Agriculture and Biological Sciences*, 3 (5): 463-473.

- Hjeljord, L. and Tronsmo, A. 1998. Trichoderma and Gliocladium. biological control: an overview. In: Trichoderma and Gliocladium: Enzymes, biological control and commercial applications. Harman GE, Kubice CP.(Eds), 2; 131-151.

- Horsfall J.G., 1945, Antagonism and Synergism. In: *Fungicides and their action,* Frans Verdoorn, Chronica Botanica Company,Waltham, Mass, (ed.), U.S.A,, 159-188.

- Jayakumar, J. and Ramakrishnan, S., 2009. Evaluation of avermectin and its combination with nematicide and bioagents against root knot nematode, *Meloidogyne incognita* in tomato, *J. Biol. Control,* 23 (3): 317–319.

- Jayaraj, J., and Ramabhadran, R., 1998. Effect of certain nitrogenous sources on the *in vitro* growth, sporulation and production of antifungal substance by *Trichoderma harzianum, Journal of Mycology and Plant Pathology,* 28 (1): 23-25.

- Kredics, L., Antal, Z., Manczinger, L., Szekeres, A., Kevei, F., and Nagy, E., 2003. Influence of environmental parameters on Trichoderma strains with biocontrol potential. Food Technol. and Biotechnol., 41(1); 37-42.

- Monte, E. 2001. Understanding Trichoderma: between biotechnology and microbial ecology. *Int. J. Microbiol.,* 4(1); 1-4.

- Mukhopadhyay, A.N.,Patal, G.J. and Brahmabatt, A., 1986, *Trichoderma harzianum* a potential biocontrol agent for tobacco damping off. *Tobacco Research,*12(16): 26-35.

- Naznin, A., Hossain, M. M., Ara, K. A., Hoque, A. and Islam, M., 2015, Influence of organic amendments and bio-control agent on yield and quality of tuberose. *Journal of Horticulture,* 2(4): 1-8.

- Pavanello, E. P., Brackmann, A., Thewes, F. R., Venturini, T. L., Weber, A. and Blume, E., 2015., Postharvest biological control of brown rot in peaches after cold storage preceded by preharvest chemical control, *Rev. Ceres, Viçosa,* 62(6); 539-545.

- Prasanna, A., R. Nargund, V., B. Bheemanna, M. and Patil, V., 2002., Compatibility of thiomethoxam with *Trichoderm a harzianum.J. Biol. Cont.*, **16** (2): 149-152.

- Prasanthi, S. K., Srikant Kulkarni, Anahosur K.H. and Kulkarni S, 2000., Management of safflower root rot caused by *Rhizoctonia bataticola* by antagonistic microorganisms. *Plant Disease Research*, **15** (2): 146-150.

- Rama Bhadra Raj, M. and Raoof, M. A., 2003, Integrated approach for the management of castor wilt (*Fusarium oxysporum* f.sp. *ricini* Nanda and Prasad). *Indian Journal of Plant Protection*, **31**(1) 64-67.

- Sharma, S. D. and Ashok Mishra, 1995., Tolerance of *Trichoderma harzianum* to agrochemicals. Indian Journal of Mycology and Plant Pathology **25** (1&2): 129.

- Shylaja, M. and Rao M. S., 2012., *In- vitro* compatibility studies of T*trichoderma harzianum* with inorganic fertilizers. *J. Nematol. medit.*, **40:** 51-54.

- Srinivas P. and Ramkrishnan G. 2002., Use of native microorganism and commonly recommended fungicides in integrated management of rice seed borne pathogens. *Annl .Pl. Protect. Sci.*, **10**(2); 260-264.

- Valarmathi, P., Kumar S.P., Vanaraj, P., Ramalingam R. and Chandrasekar G., 2013., Compatibility of copper hydroxide (kocide 3000) with biocontrol agents, *IOSR-JAVS*, **3**(6): 28-31.

- Vincent, J. M. 1947., Distortion of fungal hyphae in the presence of certain inhibitors. Nature, **159**(4051); 850-850.

- Weed, C.M., 1889, Notes on experiments with remedies for certain plant diseases. In: Ohio Agr. Exp. Sta. Bull. Series-2, 186-189.

17

Mass Production of the Nuclear Polyhedrosis Virus (NPV) of *Helicoverpa armigera* and *Spodoptera litura*

Rajeshwari R.
University of Horticulture Sciences, Bagalkot

Introduction

To feed the ever increasing population of our country like India, green revolution offered self sufficiency in agricultural production of food grains. The technological interventions have created lot of disturbances in the ecology and have prompted in the development of serious environmental problems. Newer molecules introduced for crop production such as highly persistent chemical insecticides have lead to the serious problems in man and animals by killing the beneficial flora fauna and. The pests of crops evolved rapidly with multiple insecticide resistance devastating the entire crop. Hence, eco-friendly pest management strategies such as biological control of pests using microbials are to be used which are safe with no harmful effects on environment and non target organisms. The role of entomopathogenic viruses in global crop protection has grown in the last decade and virus products are species specific. Among the different groups of entomopathogenic viruses (Eberle et al., 2012), four genera of baculoviruses, Alpha-, Beta-, Gamma-, and Delta baculoviruses (Jehle et al., 2006; Eberle et al., 2012) are important. Among these, only the lepidopteran-specific nucleopolyhedroviruses (NPV: Alphabaculovirus spp.) and granuloviruses (GV; Betabaculovirus spp.) have been commercially developed and utilized.

Nuclear polyhedrosis viruses are baculoviruses which are pathogens of insects belonging to orders namely, Lepidoptera, Hymenoptera and Diptera. The important insect viruses are *Helicoverpa armigera* NPV (HearNPV), *Spodoptera litura* NPV (SpltNPV), *Autographa californica* NPV (AucaMNPV), *Plutella xylostella* GV (PlxyGV), *Spodoptera exigua* NPV (SeMNPV) and others (Yang *et al.*, 2013).

Classification of Baculoviruses

Baculoviruses are bacilliform, enveloped viruses with covalently closed circular double-stranded DNA genomes ranging in size from 82 kb to 180 kb in size (Lauzon *et al.*, 2004) and are predicted to encode for about 90 to 180 genes. The genome is packaged in bacillus-shaped nucleocapsids and hence the name "baculovirus". Rod shaped nucleocapsid is surrounded by lipoprotein envelope referred as virion (Occlusion Derived Virus). These virions are occluded either within polyhedral or granular Occlusion Bodies (OB's). Polyhedral- shaped occlusion bodies (POB's) of NPV's range from 0.15 to 15 µm in diameter and are composed of matrix protein called polyhedron (Hooft van Iddekinge *et al.*, 1983) while GVs produce much smaller ovi-cylindrical occlusions from 0.3 to 0.5 µm in diameter (Williams and Faulkner, 1997) which are composed of a major matrix protein called granulin (Funk *et al.*, 1997; Winstanley and O'Reilly, 1999).

The family Baculoviridae was earlier subdivided into two genera, granuloviruses (GVs) and nucleopolyhedroviruses (NPVs) based on the morphology of the occlusion bodies (OBs) they form in an infected host cells. GVs are composed of one nucleocapsid per envelope, while NPVs are packaged either as one nucleocapsid per envelope (SNPVs) or many nucleocapsids per envelope (MNPVs) (Adams and McClintock, 1991). GVs have been found solely from lepidopteran hosts, whereas NPVs have been isolated from a wider range of insects. About 90 per cent of baculoviruses are reported from 34 different families within lepidoptera. The recent taxonomy proposed by Jehle *et al.* (2006) mentioned in ICTV report, divides family, baculoviridae into four genera on the basis of genome sequence analysis and its arthropod host (Herniou *et al.*, 2012), *viz.*, Alphabaculoviruses (Lepidoptera-specific NPVs), Betabaculoviruses (Lepidoptera-specific GVs), Gammabaculoviruses (Hymenoptera-specific NPVs) and Deltabaculoviruses (Diptera-specific NPVs).

Multiplication and replication cycle of Baculoviruses

Baculoviruses are obligate pathogens and require host cell for their replication. Baculovirus infection is characterized by a biphasic replication cycle where two virion phenotypes are produced and cell-to-cell spread/infection of virus is carried out by non-occluded Budded Virus (BV) whereas host-to-host transmission is achieved by the Occlusion-Derived Virus (ODV). The later are released from diseased host, spreads infection by infecting midgut epithelial cells. Occlusion bodies are ingested by insect larvae while feeding on contaminated foliage. They are solubilized in the alkaline environment by the combined action of the alkaline gut pH and proteases of the midgut and thereby releasing ODVs. Subsequently, ODVs pass through the peritrophic membrane lining the gut and fuse with the membrane of the microvilli of midgut epithelial cells which is the initial site of infection. The DNA-containing nucleocapsids are released and transported to the nucleus to initiate infection (Ohkawa et al., 2010). The viral DNA is uncoated followed by transcription and genome replication. The newly synthesized genomes are packaged and transported to the basal site of infected midgut cell to produce BVs, which spread the infection from cell to cell inside the insect body (Williams et al., 2017). Death of the host insect occurs followed by liquefaction of the internal body and weakening of the cuticle releasing progeny OBs into the environment. The virions released will be ingested by other susceptible insects.

Baculoviruses primarily infect the larval stages of insects. The progression and signs of baculovirus disease is influenced by numerous factors including the stage (instar) and physical attributes of the host insect, temperature, inoculum dose and virulence of virus. Ingestion of numerous polyhedra by early instars (1st -3rd) can cause the death within 72 hours. However, if host is infected with same amount of virus inoculum during later instars (4th or early 5th), the disease develops over a period of 5-10 days (Federici, 1997). Diseased larva becomes sluggish at 3 to 4 days after infection and begins to feed slowly. Later larvae appear swollen and cuticle looks glossy. At an advanced stage of the disease, cell lysis occurs, large numbers of infected hemocytes and polyhedra are released into the blood. The larvae dies within 1 or 2 days after infection. Before death, the infected larvae often climb to the top of vegetation and attach themselves to vegetation by their prolegs with the head and rear portion of the abdomen facing downward in an inverted 'V' shape. This is known as tree top disease. After death, the larvae lose their turgor, cuticle becomes flaccid, fragile and within few hours, the cuticle ruptures, liberating billions of polyhedra into the environment.

Isolation of NPV

The NPVs, being obligate pathogens and highly host specific, have to be isolated from their respective host insects either from laboratory colonies or from the insect populations collected from host plants in the field. The NPV disease show characteristic symptoms in host larvae like pale colouration, flaccidity and sluggish movement of infected larvae. Often, larvae killed by the virus can be seen hanging head downwards from twigs or leaves. In advanced stages of infection, the skin ruptures releasing a pale white or cream coloured fluid containing several thousands of polyhedral inclusion bodies (PIB). Microscopic examination of a smear will reveal innumerable polyhedra. Cadavers can be collected in sterile distilled water and stored at 40 °C for several years without significant loss of virulence. Solitary insects like *H. armigera* have to be collected in large numbers and reared in the laboratory under stressed conditions to bring out latent infections.

Extraction of Occlusion Bodies

To extract the OBs, 1 ml sterile distilled water was added to the cadaver and disrupted by vortexing for about 2 min. The extract was filtered through glass wool and the glass wool was washed with 500 µl of sterile distilled water and the filtrate was centrifuged at 15,000xg for 5 min. The supernatant was removed carefully, pellet was washed with 2 ml distilled water and centrifuged at 15,000xg for 5 min. The pellet was resuspended in 1 ml of sterile distilled water and stored at 4°C. The OBs were enumerated using Neubauer's haemocytometer mounted on a phase-contrast light microscope at 10 x 40 magnification. Haemocytometer is a thick glass slide with a shallow depression in the grid area at the centre. Finely ruled grid lines and squares are visible under microscope. The dimension of the grid is defined and when cover slip is placed over the grid chamber the depression of fixed depth (0.1 mm) is created. The central section of 1 mm² grid area is divided into 25 large and equal squares and area of each large square is 0.04 mm². Each large square is further divided into 16 small squares, each measuring an area of 0.0025 mm². Overall, the total grid area (1 mm²) of haemocytometer is divided into 400 smaller squares.

Enumeration of POBs:

1. Place the haemocytometer on flat, clean white surface and place the cover slip on grid line at the centre of haemocytometer and press firmly.

2. Pipette out 10 µl of diluted virus suspension into the edge of cover slip so as to spread it into the counting chamber.

3. Wait for 20 minutes to reduce the Brownian movement of POB's and observe the POB's under a compound microscope preferably a phase contrast microscope.

4. Count the number of polyhedra in each small square from at least five large squares i.e. total 80 small squares.

5. Since the depth of counting chamber, dimension of the grid area and dilution of the viral suspension used is known, the concentration of the virus is calculated using formula,

Number of polyhedra / ml = (D x X) / (N x K)
Where, D: dilution factor
X: total number of polyhedra counted
N: number of small squares considered for POB count
K: volume above one smallest square (2.5×10^{-7} cm^3)

Example: Suppose in a sample diluted 1:200 we counted 500 polyhedra in a total of 100 small squares (the area of a small square is 1/400 mm^2) using a 0.1-mm-deep haemacytometer, then: D= 200, X = 500, N = 100, K = 11(4 x 106) (i.e., 1/400 mm^2 x 0.1 mm depth x 1/1000 (1000 mm3 = 1 cm3)) thus, 200 x 500/100 x 1(4 x 106) = 200 x 5 x (4 x 106)= 4 x 109 polyhedra per ml of undiluted sample. Finally the concentration of the virus in the suspension is expressed as Number of POB's/ml.

Bioassays to test the virulence of NPVs against larvae

Bioassays are conducted on 2nd and 3rd instar larvae by providing 1.5–2.5 g diet per larva through surface contamination method. Prepare six concentrations of virus inoculums by serial dilution. The 2nd instar larvae are tested with concentrations ranging from 1.8 x 10^7 to 1.8 x 10^2 OBs/ml and 3rd instar larvae are tested with concentrations ranging from 1.8 x 10^8 to 1.8 x 10^3 OBs/ml. For each treatment, three replications are maintained with ten larvae per replication. For HearNPV and SpltNPV bioassay, dispense 50 μl of virus suspension over the surface of artificial diet in the cell wells and spread uniformly over the surface using blunt glass rod. Allow it to dry for 10–20 min and then release the larvae carefully over the surface of the diet. For AmalNPV bioassay, spray the groundnut foliage with virus for couple of times in plastic trays and allow it to dry for 10–20 min. Transfer the inoculated leaves to plastic cages and then release the larvae carefully on to the leaf surface. Rear the larvae under controlled conditions with 16 h light and 8 h dark photoperiod, 25 ± 2°C temperature and 70% relative humidity. Observations on mortality are recorded daily from fourth day after inoculation. The data are analyzed using Probit analysis software to arrive at lethal concentration of virus required to cause 50% mortality (LC50) and lethal time required to cause 50% mortality (LT50) (Chi, 1997).

Mass production and standardization of Hear NPV

Baculoviruses are obligate pathogens having high degree of host-specificity; therefore they can be mass produced only in respective host insects. Insect viruses are multiplied in-vivo on host larvae and also in-vitro on insect cell lines. The success of baculovirus as microbial control agent is governed by relative cost of production, availability of skilled manpower, adequate supply of healthy (disease free) host larvae and quality of the product. The most commonly used production method at commercial scale in our country is in-vivo replication of virus in the host insect larvae. However, in-vivo production of insect viruses in largely influenced by rearing conditions and nutrition of the host insect. Production and recovery of POB's mainly depends on the age and weight of host larvae. Quantification of POB's in a viral suspension is essential to obtain the accurate counts of infectious units (OB's) for the purpose of bioassay studies, virus propagation and field application, etc. Use of appropriate dose/inoculum of virus is highly essential to obtain better recovery of virus, accurate results of bioassay and better field performance. Haemocytometer is commonly used for counting and standardizing the POB's of insect viruses as the polyhedra are distinctive and easily visible under phase contrast microscope. The POB's are highly refractile protein crystals which can be seen as bright bodies under phase contrast illumination.

Rearing of laboratory hosts

Rearing of larvae in the natural host plant will involve frequent change of food at least once a day during the incubation period of 5-9 days increasing the handling time. In order to reduce the cost, field collected larvae are released into semi synthetic diet treated with virus inoculum. Mass culturing of insects in semi synthetic diet involves high level of expertise, hygiene and cleanliness. Collection of a large number of larvae in optimum stage (late IV / early V instars) is time-consuming and can be expensive in terms of labour and transportation costs.

Mass production of NPV on commercial scale is restricted to *in vivo* procedures in host larvae which are obtained by

» Field collection of *H. armigera and S. litura*

» Mass culturing in the laboratory in semisynthetic diet

Rearing of laboratory hosts

Rearing of larvae in the natural host plant will involve frequent change of food at least once a day during the incubation period of 5-9 days increasing the handling time. In order to reduce the cost, field collected larvae are released into semi synthetic diet treated with virus inoculum. Mass culturing of insects in semi synthetic diet involves high level of expertise, hygiene and cleanliness. Collection of a large number of larvae in optimum stage (late IV / early V instars) is time-consuming and can be expensive in terms of labour and transportation costs.

Production procedure

The NPV of *H. armigera* and *S. litura* are propagated in early fifth instar larvae of *H. armigera* and *S. litura* respectively. The virus is multiplied in a facility away from the host culture laboratory. The dose of the inoculum used is 5x105 POB in 10 ml suspension. The virus is applied on to the semisynthetic diet (lacking formaldehyde) dispensed previously in 5 ml glass vials. A blunt end polished glass rod (6 mm) is used to distribute the suspension containing the virus uniformly over the diet surface. Early fifth instar stage of larvae are released singly into the glass vials after inoculation and plugged with cotton and incubated at a constant temperature of 25°C in a laboratory incubator. When the larvae exhausted the feed, fresh untreated diet is provided. The larvae are observed for the development of virosis and the cadavers collected carefully from individual bottles starting from fifth day. Approximately, 200 cadavers are collected per sterile cheese cup (300 ml) and the contents are frozen immediately. Depending upon need, cadavers are removed from the refrigerator and thawed very rapidly by agitation in water.

Processing of NPV

The method of processing of NPV requires greater care to avoid losses during processing. The cadavers are brought to normal room temperature by repeatedly thawing the container with cadaver under running tap water. The cadavers are homogenized in sterile ice cold distilled water at the ratio 1: 2.5 (w/v) in a blender or precooled all glass pestle and mortar. The homogenate is filtered through double layered muslin cloth and repeatedly washed with distilled water. The ratio of water to be used for this purpose is 1: 7.5-12.5 (w/v) for the original weight of the cadaver processed. The left over mat on the muslin cloth is discarded and the filtrate can be semi-purified by differential centrifugation. The filtrate is centrifuged for 30-60 secs. at 500 rpm to remove debris. The supernatant is next centrifuged for 20 min at 5,000 rpm. Then the pellet containing the polyhedral occlusion bodies (POB) is suspended in sterile distilled water and washed three times by centrifuging the pellet in distilled water at low rpm followed by centrifugation at high rpm. The pellet finally collected is suspended in distilled water and made up to a known volume, which is necessary to calculate the strength of the POB in the purified suspension.

Registration

The virus formulation to be commercialized must first be registered with the Central Insecticide Board (CIB) following the stipulated protocols. The formulation should confirm to the following standards

1. The product should not contain any human or animal pathogen eg. Salmonella, Shigella, Vibrio etc.

2. The contaminant microbial population should not be more than 10^4 cfu per g/ml.

3. No contamination by chemical pesticides.

4. The LC_{50} value should not be more than the following for early II instar larvae by the diet surface contamination method.

References

- Adams, J. R. and Mc Clintock, J. T. (1991). Baculoviridae, nuclearpolyhedrosis viruses Part1. Nuclearpolyhedrosis viruses of insects. In "Atlas of Invertebrate Viruses" (Eds: Adams, J. R. and Bonami, J. R.). CRC Press, Boca Raton pp. 87-205.

- Eberle, K.E., Jehle, Johannes, Huber and Jochem (2012). Microbial control of crop pests using insect viruses. *In*: Microbial control of crop pests using insect viruses. 281-298 pp.

- Federici, B. A. (1997). Baculovirus Pathogenesis. In: The Baculoviruses (Eds: Miller, L. K.). Springer Science + Business Media, New York. pp 33-59.

- Funk, C. J., Braunagel, S. C. and Rohrmann, G.F. (1997). Baculovirus Structure. In: The Baculoviruses (Eds: Miller, L. K.). Springer Science + Business Media, New York. pp 33-59.

- Herniou, E. A., Olszewski, J. A., Cory, J. S. and O'Reilly, D.R. (2003). The genome sequence and evolution of baculoviruses. *Ann. Rev. Entomol.*, **48**:211–234.

- Hooft van Iddekinge B. J. K, Smith, G. E. and Summers, M. D. (1983). Nucleotide sequence of polyhedron gene of *Autographa californica* nuclear polyhedrosis virus. *Virology*, **131**: 5561-5565.

- Jehle, J. A., Blissard, G.W., Bonning, B.C., Cory, J.S., Herniou, E.A., Rohrmann, G. F., Theilmann, D.A., Thiem, S.M. and Vlak, J.M. (2006). On the classification and nomenclature of baculoviruses: A proposal for revision. *Archives Virology*, **151**:1257- 1266.

- Lauzon, H.A.M., Lucarotti, C.J., Krell, P.J., Feng, Q., Retnakaran, A. and Arif, B.M. (2004). Sequence and organization of the *Neodiprion lecontei* nucleopolyhedrovirus genome. *J. Virol.*, **78**:7023–7035.

- Ohkawa, T., Volkman, L.E. and Welch, M. D., 2010. Actin-based motility drives baculovirus transit to the nucleus and cell surface. *J. Cell Biol.*, **190**: 187–195.

- Williams, G. V. and Faulkner, P. (1997). Cytological changes and viral morphogenesis during baculovirus infection. In: The baculoviruses (Eds. Miller, L.). NewYork 7 Plenum. pp.61–107.

- Winstanley, D. and O'Reilly, D. (1999). Granuloviruses. In: Encyclopedia of Virology 2nd Ed. (Eds: Webster, R. G. and Granoff, A.), Academic Press, London, pp. 127–130.

- Yihua Yang, Yapeng Li and Yidong Wu (2013). Current status of insecticide resistance in *Helicoverpa armigera* after 15 years of Bt Cotton planting in China. *Journal of Economic Entomology*, **106** (1): 375–381.

18

Volatiles as Bioformulations in Agriculture

Raghavendra A.

ICAR-National Bureau of Agricultural Insect Resources, Bangalore

Introduction

The term pesticide covers a wide range of compounds including insecticides, fungicides, herbicides, rodenticides, molluscicides, nematicides, plant growth regulators and etc. The introduction of other synthetic insecticides such as – organophosphate (OP) insecticides in the 1960s, carbamates in 1970s and pyrethroids in 1980s and herbicides and fungicides in the 1970s–1980s contributed greatly to pest control and agricultural output (Wasim *et al.*, 2009). However, due to their adverse effects, the search for alternative methods to control pests and diseases is an urgent need. One of the promising products would be Bio-formulations which is a combination of bio-fertilizer and bio pesticides. A bio-fertilizer is a substance which consists of living microorganisms which, when applied to seeds, plant surfaces, or soil, colonize itself into the system and promotes growth by increasing the supply or availability of nutrients to the host plant eg: *Rhizobium, Azotobacter, Azospirilium* and Blue Green Algae (BGA) belonging to a general Cyanobacteria genus, *Nostoc* or *Anabaena*, they fix atmospheric nitrogen (Vessey, 2003). Bio pesticides are naturally existing pest control agents that are obtained from natural substances. These include microbial pesticides which control insects, weeds, fungi, and bacteria, plant pesticides which are produced from genetic materials that are added to the plant, instead of the plant growing them naturally and whereas biochemical pesticides are produced by the plants and help control pests by non-toxic mechanisms. Generally the *Bacillus* species will be used along with the Plant Incorporated Protectants (PIP). Among the entomopathogenic fungi, formulations of *Metarhizium, Beauveria, Paecilomyces, Lecanicillium, Nomurea, Aschersonia, Hirsutella* and *Entomophthora* are considered the most important genus. The mechanism of action of entomopathogenic fungi involves several processes until the insect is completely colonized and killed (Alves *et al.*, 2008). In addition to the microorganisms, the bio-pesticide consists of repellants and contain natural and plant based compounds that release odours unappealing and irritating to insects.

Volatiles in Bio Formulations

It is well documented that volatiles released from the plants or a host, and sometimes by the associated micro flora may also play a great role in host and pest interactions or otherwise known as tritropic interactions. For example, the fruit fly, *Bactrocera* species will be attracted to the variety of mango plants during the fruiting season and lay their eggs on the fruits due to the influence of volatiles. Similarly, the borers such as *Batocera rufamaculata* in mango, banana weevil *Cosmopolites sordidu*, fruit sucking moth, *Eudocima materna* in pomegranate, *Neoplocaederus ferrugineus* in cashew, Eucalyptus gall wasp, *Leptocyba invasa*, coffee white stem borer, *Xylotrechus quadripes*, paddy stem borer and rice bug, *Leptocorisa acuta*, cotton ball worm, *Helicoverpa armigera*, jute borer, *Spylozoma*, Uzi fly, *Exorista bombysis*, mulberry leaf roller, *Diaphania pulverulentalis*, borer, *Lepidiota mansueta*, sweet potato weevil, *Cylas formarcarius* etc., are greatly influenced by the volatiles released by the host plants and its surroundings. These volatiles all together can be broadly termed as semiochemicals which involves pheromones (Intraspecific) and allelochemicals (interspecific) such as Kairomone, Allomone, Synomone and Apneumone. These can be used as a trapping system for some of the agriculturally important pests especially in horticultural crops. They can be integrated with bio control agents and as a result, we can conserve the natural enemies also.

Formulation process

Extraction and analysis of these volatiles involves, a) solvent extraction: by using some of the non polar solvents *viz*: hexane, dichloromethane, pentane petroleum ether and sometimes with less or medium polar solvents such as acetone, methanol and etc. b) Dynamic head space collection of the volatiles in which the volatiles accumulated above the biological sample will be collected and directly injected into the analytical instruments for analysis and c) air entrapment method in which the purified air will be passed over the biological sample and the volatiles will be collected or trapped in an adsorbent material and later it will be eluted with a suitable solvent and subsequently processed for assay and analysis. Some of the analytical instruments used in chemo ecological studies are, Gas Chromatography coupled with Electroantennogram Detector (GCEAD), which is used to identify the time of response by an insect to the odour molecules of the host or its mating partner. Gas Chromatography with Mass Detector (GCMS) is employed for the detection or identification of the compound to which an insect has shown response at a mentioned time. It matches/compares the eluted compound with the standards stored in library. Further, the Fourier Transform Infrared Spectroscopy (FTIR) is used to detect the functional group of a compound to which the insect has shown response. Similarly, in the next step, the compound will be processed/analysed in Nuclear magnetic Resonance (NMR) for the final confirmation of the identified compound and structural elucidation of the same. The final identified compounds must be rechecked by following the same procedure mentioned above. Accuracy and specificity of the identified or analysed sample/compound depends entirely on the volatile collection, processing and purification and subsequent accurate identification in case of its molecular characterization, functional groups and matching with the standards. The presence of various compounds at different concentration in a biological extraction/sample has to be estimated and accordingly the formulation has to be prepared. Any fault in the above process will lead to misidentification of the compounds there by leading to malfunction of the entire formulations and ultimately end up in negative results in field trials. Along with these, the knowledge of pest and its host physiology, behaviour and their surrounding environment is also essential and beneficial in formulating the lures or traps. Based on several research and field experiments, several plant material based formulations have been introduced for the control of agricultural pests especially for horticultural crops.

Trapping/killing the pests

These formulations can be used for monitoring, for trapping, for mating disruption, for male annihilation, auto confusion, as a repellant etc. Once the identified biological compounds in their respective concentrations were formulated and made as lures, these in turn placed in traps which are host specific. Traps made and designed for lepidopteran pests may not be of much use in catching the coleopteran pests. Proper distance between traps, height in which they are placed, quality and quantity of the lures used should be made ascertain to ensure their effectiveness. There are several types of traps available in the market and some are customized depending upon the targeted pest behavior and its habitat. Few traps currently used in the agriculture practices are: funnel traps, water traps, bucket traps, sticky traps, delta traps, fly water traps, plastic moth traps, sleeve traps, liquid bait or fruit fly traps, light traps etc. Some of the plant compound based formulations were also prepared in the liquid form which can be sprayed, in paste form of semi solid consistency which can be swabbed to the host plants as repellants. In the following table brief information on the traps and lures used for the various types of pests are given.

Type of Traps with lures attached	Target pest	Host plants
Bucket traps	Red palm weevil *Rhynchophorus ferrugineus*	Palm trees
Yellow sticky traps	*Tuta absoluta* Hoppers of various species	Tomato Mango and some vegetable crops
Cross wan X traps	Coffee white stem borer *Xylotrechus quadripes*	Coffee
Sleeve/funnel traps	Lepidopteran pests viz: *Helicoverpa armigera, Leucinodes orbonalis* and etc.	Cotton, tomato, brinjal etc
Delta traps	Eucalyptus gall wasp *Leptocyba invasa* Fruit fly *Bactrocera* sp.	Eucalyptus Mango, guava etc
Liquibaitor fruit fly traps	Fruit fly *Bactrocera* sp. and *Cucurbitae* sp.	Mango and vegetables belonging to *Cucurbitae* family
Light traps	Various species of Cerambycidae family, *Lepidopteran* pests and etc	Forest trees, vegetable crops etc
Water traps	Root grubs and some lepidopteran moths	Varied host Mulberry plants

A brief information on use of various volatile based formulations for monitoring of insect pests in orchards, forest, field crops, male annihilation technique and mating disruption are given in the following table.

Technique	Pest, host and country	
Mating Disruption	Pink bollworm, *Pectinophora gossypiella* in cotton (USA)	
	Grape moth in grapes (Germany, France, Spain, Italy)	
	Oriental fruit moth, *Grapholitha molesta* in peaches (Australia, France, Italy, S.Africa, USA)	
	Codling moth, *Laspeyresis pomonella* in apples (Switzerland, Germany, USA, Australia)	
	Rice stem bore, *Chilo suppressalis* in rice (Spain)	
	Tea tortrix, *Homona magnanima, Adoxophytse* spp. in tea (Japan)	
	Tree borer, *Synathethedon* spp. in apricots (Japan)	
Male annihilation or mass trapping	Spruce bark beetle, *Ips typographus* in forest (Scandinavia)	
	Palm weevil, *Rhynchophorus palmarum* in oil palm (Costa Rica)	
	Japanese beetle, *Popillia japonica* in turf grass (USA)	
	Olive fly, *Dacus oleae* in olives (spain, Greece)	
	Housefly, *Musca domestica* in household (worldwide)	
	Cockroaches, *Blatella germanica, Periplaneta Americana* in household (worldwide)	
	Coco pod borer, *Conopomorpha cramerella* in cocoa (Malayasia)	
	Yellow stem borer, *Scirpophaga incertulas* in rice (India)	
Monitoring field crop pests	Diamond back moth	USA, Europe, India, SE Asia
	Pink bollworm	USA, Peru, Israel, Egypt, India
	American bollworm	India, Egypt, Pakistan, Australia
	Tobacco caterpillar	India, Japan
	Sweet potato weevil	USA
	Fruit fly	India
	Brinjal fruit & shoot borer	India
	Sugarcane early shoot borer	India
	Rice yellow stem borer	India, Pakistan, Bangladesh
Monitoring insect pests in orchards	Codling moth	USA, Europe, Asia, Australia
	Plum fruit tortrix	Europe
	Oriental fruit moth	USA, Europe, Australia and S.Africa
	Grape moth	Europe, USA
	Olive moth	Spain, Greece, Itly
	Mediterranean fruit fly	Worldwide
	Oriental fruit fly	Worldwide
	Melon fly	Worldwide
	California red scale	USA, Israel

Conclusion

Besides bioformulations, crop rotation, pruning, thinning, application of available predators, parasites, male sterilization technique, use of behavioural modifiers such as attractants, antifeedants, antioviposition molecules, repellents are other forms of methods to manage pests.

All together, bio formulations can be the combinations of the various PIP, Micro organisms, and even the plant based chemical compounds (herbal) or volatiles which has become one of the major tool in agriculture not only for controlling the pests and diseases but also to bring socioeconomic revolutions in the farming community. Based on their performance and requirements, they can be used as bio fertilizers, bio pesticides and herbal formulations exclusively in the agricultural practices with no or very less adverse effect on environment and the non targeted species.

References

• Alves, S.B., Lopes, R.B., Vieira, A.S., Tamai, M.A., 2008. Fungos entomopatogênicos usados no controle de pragas na América Latina. In: Alves SB, Lopes RB (eds). Controle Microbiano de Pragas na América Latina: avanços e desafios. 1st ed. Piracicaba: FEALQ; 2008.

• Vessey, J.k., 2003, Plant growth promoting rhizobacteria as bio-fertilizers. Plant Soil, 255: 571-586

• Md. Wasim Aktar, Dwaipayan Sengupta and Ashim Chowdhury, 2009, Impact of pesticides use in agriculture: their benefits and hazards. Interdiscip Toxicol. 2(1): 1–12. doi: 10.2478/v10102-009-0001-7

19

Isolation, Identification and Selection of Potential Antagonists

Rajeshwari R. and Vikram Appanna
University of Horticultural Sciences, Bagalkot

Introduction

Biological control has been considered a viable alternative method to manage plant diseases of horticulture crops. It is the inhibition of growth, infection or reproduction of disease causing pathogens by antagonists/bio-control agents. It is environmentally safe and offers protection to crops against diseases.

An antagonist/bio-agent should possess following characteristics

- » High rhizosphere competence

- » High competitive saprophytic ability

- » Enhanced plant growth

- » Ease for mass multiplication

- » Broad spectrum of action

- » Excellent and reliable control

- » Safe to environment

- » Compatible with other bio-agents

» Should tolerate desiccation, heat, oxidizing agents and UV

Mode of Action of antagonists

Competition: Antagonists compete with pathogens for space, minerals and organic nutrients to proliferate and survive in their natural habitats.

Antibiosis: It is an antagonistic association between two organisms mediated by specific or non-specific metabolites *viz.*, lytic enzymes, volatile compounds etc.

Mycoparasitism/Hyperparasitism: The antagonists parasitize the pathogen by coiling around the hyphae by secretion of hydrolytic enzymes such as chitinases, cellulases, glucanases etc.

1. *Trichoderma* as fungal antagonist

Trichoderma spp., a filamentous fungi, are among the microorganisms which are most commonly used as biological control agents and are marketed as biopesticides, biofertilizers, growth enhancers and stimulants of natural resistance, due to their ability to protect plants, enhance vegetative growth and check pathogen populations under varied agricultural conditions, as well as to act as soil amendments/inoculants for improvement of nutrient uptake, decomposition and biodegradation. *Trichoderma* products are promoted as biopesticides, biofungicides, bioprotectants, bio-inoculants, bio-stimulants, bio-decomposers, biofertilizers, plant growth promoters, etc. The classical mechanism of biological control by *Trichoderma* is through direct antagonism of phytopathogenic fungi by competition, antibiosis and attack with hydrolytic enzymes.

Some isolates of *Trichoderma* are well known biocontrol generalists, able to function against a broad spectrum of fungal pathogens including *Rhizoctonia, Botrytis cinerea, Sclerotinia sclerotiorum, Sclerotium* spp., *Pythium ultimum, Phytophthora* spp., *Armillaria* spp., *Fusarium oxysporum, Verticillium* spp. and *Gauemannomyces graminis.* The most commonly used species of *Trichoderma* are *T. harzianum, T. atroviride, T. asperellum, T. polysporum, T. viride*, as well as a few species that belong to the related genus of *Gliocladium*. The antagonistic potential serves as the basis for effective biological control applications of different *Trichoderma* strains as biofungicides against soil, foliar and vascular pathogens, as an alternative to chemical pesticides for treatment against a wide spectrum of plant pathogens, as well as to increase resistance to abiotic stresses. Its persistence in the soil, in rhizosphere, eventually associated as endophyte, ensures long-term advantages. *Trichoderma* is capable of systemically activating plant defense mechanisms including priming that anticipate pathogen attack.

Isolation of *Trichoderma* species

Trichoderma are commonly present in soils (rhizosphere as well as non-rhizosphere and rhizoplane), phylloplane, decaying organic matter, manure heaps and crop residues.

Collection of soil samples for *Trichoderma* isolation

Soil samples should be collected from a field where the pathogen is known to be present but disease occurrence is low. Biologically active soils that contain a diverse population of microorganisms should be sampled. Soil should be collected at a depth of 15 cm in the upper surface including rhizosphere and rhizoplane samples. The collected samples should be analyzed as soon as possible. For isolation of phylloplane microflora, healthy foliar parts (like flowers, fruits, leaves) should be collected.

Composition of *Trichoderma* selective media (TSM)

Ingredients	Quantity
a) $MgSO_4.7H_2O$	: 0.2 g
b) K_2HPO_4	: 0.9 g
c) KCl	: 0.15 g
d) NH_4NO_3	: 3.0 g
e) Chloromphenicol	: 0.25 g
f) PCNB	: 0.2 g
g) Rose Bengal	: 0.15 g
h) Captan	: 0.2 g
i) Propamocarb or	: 1.2 ml
Metalaxyl	: 1.6 g
j) Water	: 1000 ml

Estimation of *Trichoderma* population in soils

Collect the soil sample from the field, mix well and make in to fine particles. Collection of soil sample should be made in the root zone at 5-15 cm depth preferably from rhizosphere.

Materials

    i.   Sterilized water, 100 ml in conical flask (250ml)

    ii. Sterilized water, 9 ml in test tubes

    iii. Sterilized pipettes 1 ml

    iv. Sterilized petri dishes

    v.   *Trichoderma* selective media (TSM)

Procedure

1. Suspend 1 g of soil sample in 9 ml of sterile distilled water (1:10 or 10^{-1}). Shake well.

2. Take 1 ml from this and transfer to 9 ml of sterile water in tube (1:100 or 10^{-2}).

3. Make serial dilutions by transferring 1 ml of the suspension to the subsequent tubes to get 10^{-3} to 10^{-8}.

4. Transfer 1 ml of each dilution to petri plates containing TSM. Rotate the plate gently. Incubate at 28°C for 5-7 days. Observe the development of colonies.

Observations

1. *Trichoderma* colonies on the selective media will be white initially and later on turn to green.

2. Count the number of colonies in each petri plates.

Calculations (for quantitative estimation)

1. Assume average number of *Trichoderma* colonies per plate is 4 at 10^{-4} dilution (1:10,000).

2. 1 ml of 1:10,000 dilution contains 4 colonies

3. 1 gm of soil will contain 4x10,000=40,000 colonies=4x10^4 colonies

Isolation from rhizosphere (dilution plate method)

1. Separate rhizosphere soil from 5-6 roots with the help of a brush in a petri plate. Add 10g of rhizosphere soil in 100 ml sterile water and shake it for 15 min on a magnetic shaker.

2. Prepare serial dilutions, 10^{-2} to 10^{-6}

3. Transfer one ml each of dilution 10^{-4} to 10^{-6} to sterile petri plates

4. Pour melted and cooled *Trichoderma* specific agar medium

5. Incubate the plates at 25ºC for 5-7 days.

6. Transfer isolated colonies on PDA slants.

Isolation from phylloplane

1. Collect fresh, healthy leaves (of all ages) in sterile polythene bags

2. Transfer 10 g to 100 ml sterile water and stir for 20 min using magnetic stirrer

3. Prepare serial dilutions (10^{-2} and 10^{-3})

4. Transfer 1ml aliquot from all dilutions to sterile petri plates and pour *Trichoderma*specific agar medium

5. Incubate the plates for 5-7 days at 25ºC for obtaining colonies of *Trichoderma*.

Phenotypic characterization of *Trichoderma*

To tentatively identify the culture of *Trichoderma* by morphology with the identification key, following steps are followed.

Procedure

Observe and analyze the cultures with naked eye and microscope. Analyze the following characters and interpret your results by comparing with the identification key provided.

Traits to be observed

Sl. No.	Trait/Morphological character to be observed	Interpretation
1.	Colony morphology	
2.	Colony colour	
3.	Growth rate	
4.	Branching pattern	
5.	Conidia-shape and size	
6.	Conidia-smooth or hairy	
7.	Phialides	
8.	Chlamydospore	

Observe the above mentioned traits and compare with the key provided (*Trichoderma* – Identification and agricultural applications-Gery, J. Samuels, APS Press, USA).

Identification of *Trichoderma* species

Trichoderma is a genus of filamentous Deuteromycetes. Various species under these genera can be identified by morphological characters. Morphological characters used in specific recognition in *Trichoderma* is primarily based on the angle at which branches and phialides are borne. Macromolecular approaches (enzyme and nucleic acid analyses) have been given importance recently because of lot of confusion in taxonomy of *Trichoderma*.

i.   *Trichoderma viridae* Pers.: Fr.

Conidia: Single-celled, green or brownish, sub-globose to ellipsoid, rough-spored.

Conidiophores: Narrow, flexuous, primary branching at regular intervals, paired or in whorls of three, short and not extensively branched.

Phialides: Mostly in virticils of 2 or 3.

Chlamydospores: Present in most isolates, globose, rarely ellipsoid, intercalary.

Colony characters: Grows rapidly, 5-9 cm diameter after 4 days at 20°C. Surface smooth, become hairy, dark green and typical coconut odour is emitted.

Reverse pigment: Colourless to dull yellowish.

ii. *T. harzianum* Rifai

Conidia: Subglobose to obvoid or short ellipsoid, 1.7-3.2x1.3-2.5 µm (avg. 2.4x1-9 µm)

Conidiophores: Hyaline, smooth walled, straight or flexuous up to 8 µm wide near base, 2.5-4.5 µm wide, most of the length, highly branched, primary branching at right angles, whorls of 2 or 3, secondary branching in whorls of 2-4, ultimate branch 1-celled, 3.5-10x2.5-6 µm.

Phialides: Ampuliform to sub-globose or lageniform, 3.5-7.5x2.5-3.8 µm, arise mostly in crowded and diverse whorls of 2-6.

Chlamydospores: Fairly abundant, intercalary and terminal, 4-12 µm. Mostly globose, smooth and intercalary.

Colony characters: Grows rapidly, floccose aerial mycelium, whitish green, bright green to dull green.

Reverse pigment: Colorless to dull yellowish

iii. *T. hamatum* (Bon.) Bain.

Conidia: Oblong to ellipsoid, dull green, 3-4.5x2.1-2.8 µm (avg. 3.9-2.5 µm).

Conidiophores: 5-10 µm diameter near the base, highly branched, primary branches relatively short, usually in whorls of 2-5, highly re-branched, ultimate branch one celled, barrel-shaped or short cylindrical, mostly 3.5-7x3.5 µm.

Phialides: subglobose to ellipsoidal or ampulliform, 3.3-5.6x2.8-3.5 µm, arise in crowded whorls of 3-6.

Chlamydospores: abundant, 33x18 µm.

Colony characters: colonies grow moderately rapidly, limited aerial mycelium is floccose, white to grayish.

Reverse pigment: Colorless to pale-greenish yellow.

iv. *T. virens* (Miller, Giddens &Foster) von Arx

Conidia: Broadly ellipsoidal to obvoid, 3.5-6.0x2.8-4.1 µm, dark green, conidia from adjacent phialides often coalescing into large globoid masses.

Conidiophores: Conidiophores sub hyaline, 30-300 µm long, 2.5-4.5 µm in diameter, towards base frequently unbranched for about half of the length, towards the apex branching irregularly with each branch terminated by a cluster of 3-6 closely addressed phialides, branches arise at right angles or refluxed towards the apex, primary branches arising singly or in opposite pairs immediately beneath the septa, re-branched irregularly once or twice, ultimate branches one or two celled.

Phialides: Ampulliform to lageniform, 4.5-10x2.8-5.5 µm, swelling in the middle, mostly arising in closely appressed verticils of 2-5 or terminal branches.

Colony characters: grows rapidly, aerial mycelium floccose, white to grayish

Reverse pigment: colourless

Identification services in India

>> Division of Plant Pathology, Indian Agricultural Research Institute, Pusa, New Delhi

>> Fungal Identification Service, Agharkar Research Institute, Pune

>> Institute of Microbial Technology, Chandigarh

>> Molecular fungal identification services, Indian Institute of Horticultural Research, Bangalore

In vitro evaluation of fungi for antagonistic effect on plant pathogens

Dual culture method

1. Six mm discs of pathogen and antagonistic fungus from 5-7 days old cultures are placed on PDA plates opposite to each other at the corners of plate.

2. Inoculate plates with either the pathogen alone or test fungus alone in a similar way serving as control.

3. After 5-7 days of incubation, measure radial growths of pathogen and test

fungus in dual and monoculture. Calculate the percentage of inhibition of pathogen/test fungus.

4. Re-isolate pathogen/test fungus (viability test) from the interaction sites by transferring 6 mm discs cut from the interaction zone and examine the renewed growth.

5. The interactions such as mutual inhibition with pigmental band, with a clear zone or extreme inhibition of pathogen occurs.

Observations to be recorded

Measure the mycelia inhibition of pathogen in millimeters when the fungus fully grown in the control plates (9 cm). Calculate the Percent Inhibition (PI) over control using the formula:

$$PI = \frac{C-T}{C} \times 100$$

Where,

C is the growth of test pathogen (mm) in the absence of the antagonist strain

T is the growth of test pathogen (mm) in the presence of the antagonist strain

Antibiosis test for production of diffusible inhibitory metabolites

1. Cut cellophane paper (50 µm) discs of 90 mm diameter and sterilize in an autoclave at 121°C for 15 min and then place each sterilized disc aseptically over potato dextrose agar medium in a petri dish.

2. Place 6 mm discs of each candidate fungus at the centre on the cellophane paper and incubate for 2-5 days.

3. Remove the cellophane paper and the adhering fungus carefully and place 6 mm disc of pathogen at the centre of the medium previously occupied by the candidate fungus.

4. Record radial growth of pathogen every 24 h and compare with the growth in control plates.

5. Calculate the percentage inhibition of pathogen.

Antibiosis test for production of volatile compounds

Many of the antagonistic organisms produce chemicals that are inhibitory to the pathogens. These compounds may either be volatile or be released into the medium (nonvolatile).

1. Seal 2 petri dish bases (90 mm) containing 20 ml of solidified potato dextrose agar medium together by an adhesive tape and grow antagonistic fungus in the bottom plate for 1, 3, 6 and 10 days at 25±2 °C and allow the medium on the upper plate to adsorb any volatile compounds produced by the candidate fungus.

2. Remove upper potato dextrose agar plates after the required exposure to the candidate fungus and inoculate with pathogen.

3. Record radial growth of pathogen after every 24 h and compare with the growth in control plates.

4. Calculate percentage inhibition of pathogen.

2. *Pseudomonas* as bacterial antagonist

Pseudomonas fluorescens is a plant growth promoting rhizobacteria (PGPR) which colonize the root system of plants and can stimulate plant growth by direct or indirect mechanisms. Direct mechanisms of plant growth promotion include biofertilization, stimulation of root growth, rhizo-remediation and plant stress control, while mechanisms of biological control include reducing the disease incidence, antibiosis, induction of systemic resistance, competition for nutrients and niches. *Pseudomonas* has a great scope and marketing potential.

Isolation of *Pseudomonas fluorescens*

The zones from where antagonistic bacteria can be isolated are (a) rhizosphere and (b) phyllosphere or phylloplane

Composition of King's B medium for isolation of *P. fluorescens*

Ingredients	Quantity
Proteose Peptone	20.0g
Glycerol	10.0 ml
K_2HPO_4	1.5 g
$MgSO_4.7H_2O$	1.5 g
Agar	15.0 g
Distilled water	1000 ml
pH	7.0 – 7.2

Isolation from soil

1. Collect the soil samples from rhizosphere (suppressive soils) to a depth of 15-20 cm from a minimum of 5-20 points along a diagonal transect.

2. Weigh representative soil sample of 1 g from the composite soil sample collected from the field.

3. Add 1 g of the soil sample to 9 ml water blank to make 1:10 dilution (10^{-1}) and shake well on a magnetic stirrer.

4. Transfer 1 ml suspension to another 9 ml sterile water blank using another sterile pipette (10^{-2}).

5. Make further serial dilutions up to 10^{-5} by intermittent shaking.

6. Transfer 1 ml of aliquot from 10^{-5} dilution to petri plates containing King's B medium.

7. Numerous colonies of bacteria will emerge on the surface of the media. Isolate single colonies and transfer a loop full onto slants for purification. Select colonies based on colour and morphology, ie., greenish blue or dark colonies are mostly fluorescent *Pseudomonads*.

8. For long term storage of bacteria, mix the bacterial broth culture with glycerol (1:1) in small vials or tubes and keep at -80°C.

9. Screen the isolated bacteria for antagonism against plant pathogens.

Isolation from rhizosphere

1. Collect whole roots (along with closely adhering soil) from healthy plants growing in soil.

2. Place whole roots in 10 ml sterile water and dislodge rhizhosphere associated bacteria by continuous stirring for about 30 min.

3. Perform serial dilutions of the water (up to 10^{-5}) and plate out the dilutions onto the King's B medium.

4. Incubate the petri plates for 48 h at room temperature at 28 ºC.

5. Screen the isolated bacteria for antagonism against plant pathogens.

Isolation from phyllosphere or phylloplane

The bacteria are isolated from the surface of the healthy leaves.

1. Select healthy leaves of all stages from plants growing in disease prone soils.

2. Gently rinse the leaves in sterile water for few seconds to remove dust particles and loosely adhering unassociated microbes.

3. Cut five discs each of 6 mm diameter from every leaf using sterile cork borer.

4. Transfer 10 discs to 10 ml water blank and stir it for 20 min using magnetic stirrer.

5. Transfer 1 ml of the suspension to 9 ml sterile water blank using another sterile pipette diluting the original sample to 10 times (10^{-1}).

6. Shake the contents for 2-3 min.

7. Transfer 1 ml suspension to another 9 ml sterile water blank using another sterile pipette (10^{-2}).

8. Make further serial dilutions up to 10^{-5} by intermittent shaking.

9. Transfer 1 ml of aliquot from 10^{-5} dilution to petri plates containing King's B medium.

10. Alternatively, the leaf/leaves can be directly imprinted onto the solidified agar media by using both the leaf surfaces.

11. Incubate the plates for 48 h at room temperature.

12. Transfer one representative of each colony type to fresh media plates to establish pure cultures.

13. Screen for antagonism against plant pathogens.

Identification of *P. fluorescens*

Pseudomonas fluorescens is a common gram –ve, rod shaped bacterium. It belongs to family *Pseudomonadaceae* of class Proteobacteria. Cells have multitrichous flagella and are common inhabitants of soil rhizosphere, plant debris etc. They are efficient producers of potent antibiotics. The presence of bacteria is favoured by moist soil high in organic matter. The action of fluorescent *Pseudomonads* in plant growth enhancement is due to production of siderophores and pseudobactin that deprives iron availability to pathogens and there by promotes plant growth.

Identification is carried out as per Bergey's Manual of Systematic Bacteriology (8th edition, 1984). Fluorescent *Pseudomonads* produce a fluorescent pigment on King's B medium. The bacteria shows positive reaction for Kovac's oxidase test, arginine di hydrolase and gelatin liquefaction test. Identification can also be carried out at the Microbial Type Culture Collection Centre (MTCC), Institute of Microbial Technology, Chandigarh. The Gram stain is the important procedure to be followed before proceeding with the identification. The bacteria are grouped into two major groups: 1) Gram positive and 2) Gram Negative.

Procedure for Gram staining

1. Prepare bacterial smear on a clean glass slide.

2. Heat fix cells onto a glass slide

3. Stain with crystal violet for 20 seconds.

4. Mordant with Gram's iodine for 1 minute.

5. Wash or decolorize with 90% alcohol.

6. Counter stain with saffranin for 20 seconds.

7. Wash with distilled water.

8. Gram positive cells appears purple and Gram negative cells appears pink or red under the microscope.

9. Appearance of pink coloured cells confirms the presence of *P. fluorescens* which is gram negative.

Selection of potential bacterial antagonist

Three methods viz., (a) Dual culture technique, (b) Green house inhibition of plant pathogens and (c) Testing for plant growth promoting ability are used.

Dual Culture Technique

a. Streak the antagonistic bacteria in a 4 cm line on test medium.

b. Inoculate bacterial pathogen at 4 cm apart from bacterial streak.

c. Incubate plates at 28 °C and record colony growth after 8 days from 3 replicaplates.

d. Calculate the percentage of inhibition.

Filter disc method

a. Prepare suspension of pathogen as well as bacterial antagonist from 48 h old cultures.

b. Inoculate the pathogen suspension first on to nutrient agar plates by spread plate.

c. Soak 5 mm diameter filter paper disks with 20 µl of the antagonist suspension (10^8 cfu/ml)) for five minutes.

d. Place disc in the petri plates immediately after soaking.

e. Measure the inhibition zone around the filter discs 48 h after incubation.

Green house test

This test is conducted after initial screening by dual culture to confirm the inhibiting ability of the promising bacterial antagonist under potted conditions. Three methods are followed.

    a.   Root dip

    b.   Seed treatment

    c.   Soil treatment

Testing for plant growth promoting ability

The promising bacterial antagonist (identified based on inhibition zone in dual culture) are screened for plant growth promoting ability by comparing the growth parameters such as root length, shoot length, total seedling length, leaf area, fresh weight etc., of treated plants with that of control or untreated plants.

Procedure

1. Raise the seedlings from the seeds, which are treated with antagonist.

2. Grow the test plants without treating with antagonist.

3. Allow the plants to grow for a period of 30-35 days and record growth parameter such as root length, shoot length, leaf area and fresh weight.

4. Compare the observations with that of untreated plants.

20

Mass Production and Formulation of Antagonists

Rajeshwari R. and Vikram Appanna
University of Horticultural Sciences, Bagalkot

1. Mass production of *Trichoderma* spp.

There are two methods of inoculum production of *Trichoderma* spp. *viz.*, Solid state fermentation and Liquid state fermentation.

Solid state fermentation

The fungus is grown on various cheap cereal grains (sorghum, ragi and jowar), agriculture wastes and byproducts (wheat bran, saw dust, rice bran, tapioca refuse, press mud, coffee-berry husk, spent tea leaf waste, coconut coir pith, groundnut shell and poultry manure).

Preparation and sterilization of solid medium (cheap grains like, sorghum, maize, ragi or rice bran).

Material

1. Sorghum/ragi/jowar grains

2. PVP pipes of 2" length

3. Polypropylene bags or heat resistant plastic bags (25x35cm or 15x25cm size).

Technique

1. Fill polypropylene bags (25x35 cm) with 500 g of soaked sorghum/ragi grains.

2. Soak the cereal grains in water overnight and then decant water in the morning.

3. Sterilize the soaked grains for 30 min in an autoclave or pressure cooker.

4. *Trichoderma* mother culture (10 day old *Trichoderma* culture grown on potato dextrose agar) should be used for inoculation.

5. Inoculate the sterilized grains with *Trichoderma* spore suspension grown in potato dextrose broth (5 ml), plug the bags, mix the inoculated culture bag gently and incubate for 10-15 days.

6. *Trichoderma* produces dark green spores and coats on the grains. These grains are used for preparation of formulation.

Talc based formulation

Ingredients

Trichoderma culture biomass   1 L
Talc (300 mesh, white colour)   2 Kg
Carboxymethyl cellulose (CMC) or Arabic gum powder 10.0 g

Procedure for preparation

1. Talc powder of 300 mesh size procured from the authorized manufacturer is sterilized in an autoclave at 120 °C and 15 lb pressure.

2. The fungal antagonist, Trichoderma is grown in Potato Dextrose Agar (PDA). The mycelial discs grown in media are transferred to 1 kg jowar or ragi. which is soaked and autoclaved.

3. Trichoderma mycelia and spores grown in the solid medium is homogenized in a blender in to fine powder is mixed in a container with talc powder @ 1:5 w/w. The spore load of the solid ragi culture is assessed to be 2×10^6 spores /g using a haemocytometer.

4. To the above mixtures, adjuvant, Carboxy Methyl Cellulose (CMC) is added @ 5g/kg of the formulation and shade dried to 8 % moisture level.

5. The spore load of the talc formulation from liquid culture/solid substrates should be assessed to be not <2x106 cfu/g of the formulation.

6. The product is packed in polythene bags of 1kg capacity.

Liquid state fermentation

The fungus is grown in media like, potato dextrose broth, molasses yeast broth etc. *Trichoderma* can be grown in any one of the liquid media under stationary/shaker/ fermentor culture conditions. In stationary culture conditions, it will take about 10 days for full spore production. In shaker, it takes 7 days and in fermentor, peak production occurs after 3 days. Biomass from the liquid fermentation either can be separated from medium and concentrated or entire biomass with medium can be incorporated into carrier materials.

Liquid media for mass production

i) Potato dextrose/Jaggery medium

Potato	200g
Dextrose/Jaggery	20g
Water	1000 ml

ii) Molasses yeast medium

Molasses	30g
Yeast granules	5 g
Water	100 ml

iii) Jaggery soya flour medium

Jaggery	50 g
Soya flour	10 g
Water	1000 ml

vii. Formulation of *Trichoderma* spp.

Management of plant diseases depends on the development of commercial formulations with suitable carriers that support the survival of *Trichoderma* for a considerable length of time.

Characteristics of an ideal formulation

1. Should have increased shelf life.

2. Should not be phytotoxic to the crop plants.

3. Should tolerate adverse environmental conditions.

4. Should be cost effective and should give reliable control of plant diseases.

5. Should dissolve well in water.

6. Carriers must be cheap and readily available for formulation development.

7. Should be compatible with other agrochemicals.

Talc based formulation

Ingredients

Trichoderma culture biomass	1 L
Talc (300 mesh, white colour)	2 Kg
Carboxymethyl cellulose (CMC) or Arabic gum powder	10.0 g

Procedure for preparation

1. Talc powder of 300 mesh size procured from the authorized manufacturer is sterilized in an autoclave at 120 °C and 15 lb pressure.

2. The fungal antagonist, *Trichoderma* is grown in Potato Dextrose Agar (PDA). The mycelial discs grown in media are transferred to 5 ltr conical flasks each containing 3 ltr of sterilized PD medium.

3. *Trichoderma* mycelia and spores grown in the liquid medium is homogenized and is mixed in a container with talc powder @ 1:2 v/w. The spore load of the broth culture is assessed to be 10^8-10^9 spores /ml using a haemocytometer.

4. *Trichoderma* multiplied on solid substrates (jowar and ragi) are homogenized in a blender in to fine powder mixed with talc powder @ 1:5 w/w.

5. To the above mixtures, adjuvant, Carboxy Methyl Cellulose (CMC) is added @ 5g/kg of the formulation and shade dried to 8 % moisture level.

6. The spore load of the talc formulation from liquid culture/solid substrates should be assessed to be not <$2x10^6$ cfu/g of the formulation.

7. The product is packed in polythene bags of 1kg capacity.

8. The talc formulations of *Trichoderma* has been proposed to have a shelf life of 3 to 4 months.

Vermiculite-wheat bran based formulation

Ingredients	Quantity
Vermiculite	100 g
Wheat bran	33 g
Wet fermentor biomass (FB)	20 g
0.05N HCl	175 ml

Protocol: *Trichoderma* is multiplied in molasses-yeast medium for 10 days. Vermiculite and wheat bran are sterilized in an oven at 70 ºC for 3 days. Then 20 g of fermentor biomass (FB) and 0.05N HCl are added, mixed well and dried in shade.

Pesta granules (wheat bran) based formulation

Ingredients	Quantity
Wheat flour	100 g
Fermentor biomass	52 ml
Sterile water	sufficient enough to form dough

Protocol: 52 ml of FB is added to wheat flour (100 g) and mixed by gloved hands to form cohesive dough. The dough is kneaded, pressed flat and folded by hand several times. Then 1mm thick sheet (pesta) is prepared and air-dried till it breaks crisply. After drying, dough sheet was ground and passed through a 18 mesh (1.0 mm) sieve and granules were collected.

Alginate pills based formulation:

Ingredients	Quantity
Sodium alginate	25 g
Wheat flour	50 g
Fermentor biomass	200 ml

Protocol: Sodium alginate is dissolved in one portion of distilled water (25 g/750 ml) and food base is suspended in another portion (50 g/250ml). These preparations are autoclaved and blended together with biomass. The mixture is added drop wise into $CaCl_2$ solution to form spherical beads, which are air-dried and stored at 5 °C.

Press mud based formulation

Press mud is available as a byproduct of the sugar factory. The method involved is uniformly mixing of 9 days old culture of *Trichoderma* (prepared in potato dextrose broth) into 120 kg press mud. Water is sprinkled intermittently to keep it moist. This is covered by gunny bags under shade to permit air movement and trap moisture. Within 25 days, nucleus culture for further multiplication becomes ready. Press mud can also be used for mass multiplication of *Trichoderma*. The nucleus culture is added to 8 tons of press mud, mixed thoroughly and incubated for 8 days under shade condition before being applied in the field.

Coffee husk based formulation

Trichoderma formulation based on coffee husk which is a waste from coffee curing industry can be made. This product is very effective in managing *Phytophthora* foot rot of black pepper and is widely used in Karnataka and Kerala.

Invert emulsion formulations

These formulations are prepared by mixing the conidia harvested from the solid state/liquid state fermentation with a combination of vegetable/mineral oils in stable emulsion formulation. In such formulations, microbial agents are suspended in a water immiscible solvent such as a petroleum fraction (diesel, mineral oils), and vegetable oils (groundnut etc.) with the aid of a surface active agent. This can be dispersed in water to form a stable emulsion. The oils used should not have toxicity to the fungal spores, plants, humans and animals. Invert emulsion formulations/oil-based formulations are supposed to be suitable for foliar sprays under dry weather and to have prolonged shelf life.

Banana waste based formulations

A pit of different banana waste *viz.*, sheath pseudo stem and core is chopped to the length of 5 to 8 cm. A pit is prepared and different ingredients are placed in five different layers. Each layer contains one ton banana waste, 5 kg urea, 125 kg rock phosphate and one liter broth culture of *Trichoderma*. Five different layers are prepared similarly and mixed thoroughly. Banana waste is decomposed within 45 days and enriched culture is available for field application.

2. Mass production of *Pseudomonas fluorescens*

Mass production of antagonistic bacteria involves growing the microorganisms in the appropriate medium using rotary shakers or fermentors. Bacteria requires adequate nutrient supplements for proper growth.

Usually King's B medium (broth) is ideal for mass production of antagonistic bacteria, however other media like Tryptic Soya Broth or Nutrient Broth can also be used. It is better to choose the medium that provides optimum growth.

Shaker culture production of bacterial antagonists

i) King's B Medium

Ingredients	Quantity
Proteose Peptone	20.0 g
Glycerol	10.0 ml
K_2HPO_4	1.5 g
$MgSO_4.7H_2O$	1.5 g
pH	7.0-7.2
Distilled water	1000 ml

ii) Tryptic Soya Broth

Ingredients	Quantity
Soyabean casein enzymic hydrolysate	15.0 g
Papaic digest of soyabean meal	5.0 g
NaCl	5.0 g
Distilled water	1000 ml
pH	7.0 -7.2

iii) Nutrient Broth (NB)

Ingredients	Quantity
Beef Extract	3.0 g
Bacto-peptone	5.0 g
Glucose	5.0 g
NaCl	5.0 g
Distilled water	100 ml
pH	7.0-7.2

The bacteria are multiplied usually using rotary shakers in large flasks (5 lts.) under room temperature for 48 h. A loop full of 48 h old bacterial culture is used for inoculation of 50 ml of NB medium and incubated for 48 h. The primary culture is inoculated to 3 ltr media flasks and incubated for 48 h. After 2-3 days, bacterial count is estimated to be $2x10^6$ cfu/ml of the media and is used for formulation preparation. The incubation time can be chosen based on growth curve of the selected antagonist. The harvested bacteria should always be in the active phase.

Fermentor production of bacterial antagonists

For large scale mass multiplication of bacteria (industrial use), fermentors with large volume capacity can be employed. There are two types of fermentors, (a) Batch type and (b) continuous type.

a. **Batch type fermentors**: The organism is grown in the medium for a definite period of time. Then, the cell mass is separated or concentrated.

b. **Continuous type fermentors**: The inflow of the ingredients into the medium is regulated with the withdrawal of fermented product of cells in such a manner as to maintain the fermentation process continuously at a given level of operation.

All other parameters such as inoculum load, pH and temperature of the substrate are regulated so as to obtain the optimum levels of fermentation continuously. Raw materials for fermentors are sugar beet, potato, sweet potato, tapioca, apples, raisins, grapes, sweet corn, rice, blackstrap molasses, sorghum, wheat, barley, malted cereals, honey maple etc.

viii. Formulation of *Pseudomonas fluorescens*

Talc is usually used as carrier. The inoculums culture of the bacterial antagonist (containing a minimum population of 2×10^8 cfu/ml) is mixed with the sterile carrier (400 ml/kg) and air dried. Ten grams of carboxymethyl cellulose (CMC) per kg of carrier is used as adherent.

ix. Quality standards and improvement of shelf life of biopesticide formulation

Supply of quality pesticides including biopesticides/antagonistic organisms is of utmost importance for sustaining crop production from the possible damages to crops by diseases. The Central Government, while granting registration under the Insecticides Act, 1968 prescribed various parameters for monitoring the quality of pesticides including biopesticides. For implementing the various provisions of the said Act, States/Union Territories have notified four important functionaries *viz.*, licensing officer, Appellate Authority, Insecticides Analysts and Insecticides Inspectors. The notified functionaries draw samples from the manufacturing units and distribution sale points and get them analyzed/tested in the pesticide testing laboratories by the notified insecticides analysts. At present, there are 45 state pesticide testing laboratories in 18 states and in one UT with a capacity to analyse 56,000 samples per annum. There are also two Central Regional Pesticide Testing Laboratories at Chandigarh and Kanpur and one Central Insecticides Laboratory at Faridabad for quality testing of pesticides.

Quality standards of formulations

1. Colony forming units (CFU) on selective medium should be minimum of 2×10^6 CFU per ml. or gram.

2. Pathogenic contaminants should not be present. Other microbial contaminants not to exceed 1×10^4 cfu/g.

3. Maximum moisture content should not be more than 8 % for dry formulation of fungi and 12 % for bacteria.

4. Stability of the formulation at 30 ºC and 65 % RH.

Estimation of moisture of talc formulation

a. **Oven drying method:** Weigh 10 g of talc formulation and dry it in an oven at 40 °C until a constant weight is reached. Estimate the percent of moisture based on the difference in the initial weight and final constant weight.

b. **Moisture analyzer:** Electronic digital moisture analyzers are available, which can be used for moisture analysis.

Estimation of bioagent population in talc formulation

Fixed standards: CFU on *Trichoderma* selective medium should be minimum of 2×10^6 CFU per ml or g of formulation for *Trichoderma*. For formulations of *P. fluorescens*, CFU count is estimated using King's B medium by serial dilution method. The CFU should be minimum of 1×10^8 /ml or gram of formulation.

Procedure

1. Suspend 1g of formulation sample in 10 ml of sterile distilled water (1:10 or 10^{-1}). Shake well.

2. Take 1 ml from this and transfer to 9 ml of sterile water in tube (1:100 or 10^{-2}).

3. Make serial dilutions by transferring 1 ml of the suspension to the subsequent tubes to get 10^{-3} to 10^{-10}.

4. Transfer 1 ml. of the desired suspension to petri plates containing specific medium for *Trichoderma* and *Pseudomonas*.

5. Rotate the plate gently. Incubate at room temperature. Observe the development of colonies.

Observations

After 5-7 days, count the number of colonies developing in individual plates. Take average of three replications.

Calculations (for quantitative estimation)

1. Assume average number of colonies of bioagent per plate is 4 at 10^{-6} dilution (1:10,00,000).

2. 1 ml of 1:10,00,000 contains : 4 colonies

3. 1 g of formulation will contain: 4x10,00,000 colonies=4x10^6 colonies/g of formulation.

Estimation of other microbial contamination in formulations

For estimation of contamination levels (other fungal and bacterial contaminants) in the formulations, serial dilution method as described above on PDA (fungal contamination) and nutrient agar medium (bacterial contamination) may be used. Total microbial contaminants including fungal and bacterial contaminants should not exceed 1x10^4 cfu/g of formulation.

Improvement of shelf life of bioagent formulations

Shelf life of the formulated product of a biocontrol agent plays a significant role in successful marketing. In general, the antagonists multiplied in an organic food base have longer shelf life than the inert or inorganic food bases. Shelf life of *Trichoderma* in coffee husk was more than 18 months. Talc based formulation of *Trichoderma* and *Pseudomonas* have a shelf life of 3 to 4 months. For, storage of formulations polythene bags of 100 micron thickness are ideal for packing. The shelf life of talc formulations of bioagent can be increased using various ingredients (chitin and glycerol) in production medium. Heat shock at the end of log phase of fermentation extends the shelf life of talc formulation successfully up to one year.

Role of adjuvants, spreaders and stickers in formulations

Performance of *Trichoderma* and *Pseudomonas* in the formulations can be increased by the incorporation of water-soluble adjuvants, oils, stickers and emulsions. It increases the efficacy of biocontrol agents by supplying nutrients and by protecting the microbes from desiccation and death. Incorporation of carboxy methyl cellulose in formulations serves as stickers in uniform seed coating of microbes. Though adjuvants and stickers increase the efficacy of bio-products, it has its own demerits. Hence, a comprehensive knowledge on the usage of adjuvants, stickers is crucial for increasing the efficacy of formulations.

21

Isolation, Culturing, Mass Production and Formulation of Entomopathogenic Nematodes

Rajeshwari R. and Vikram Appanna
University of Horticultural Sciences, Bagalkot

Isolation and Culturing of Entomopathogenic Nematodes

Introduction

Entomopathogenic nematodes of families Steinernematidae and Heterorhabditidae are unique in their action and potential. They are considered as one of the most suitable entomopathogens in managing wide variety of insect pests particularly soil inhabiting ones. The uniqueness stems out from their symbiotic association with entomopathogenic bacteria, ease of mass production, storage and application. Due to their safety to non-target organisms and the environment, they are even exempted from Environment Protection Act in many countries. Indian research on EPNs dates back to 1960s with use of DD-136 (a commercial product of *Steinernema carpocapsae*) against several Lepidoteran pests.

It can be isolated from soil or directly collected from infected hosts, either naturally occurring or purposely deployed "trap" insects. Soil samples can be collected from different habitats, altitudes and rainfall by random sampling technique. The greater wax moth, *Galleria mellonella* is quite susceptible to *Steinernematids* and *Heterorhabditids* and makes an excellent trap insect for baiting as it is easily reared and widely available.

Isolation of EPN

Method 1: Isolation of EPN through Baiting technique

1. Take a small plastic vial of 50 ml capacity. Cut open the top and bottom portion and fix fine wire mesh (sufficient to prevent the entry of predatory ants and escape of bait larvae) to both ends or simply make fine perforations on the body of the vial.

2. Place a thin layer of filter paper along the inner wall of the vial and moist with distilled water.

3. Release four grown up last instar wax moth larvae into the vial and close it.

4. Tie a metal wire to the vial with a tag on the free end for easy identification of the vial in the field.

5. Place the vial in a suitable soil patch at a depth of 10-15 cm and close it. The free end of the metal wire along with the tag should be visible above the ground. Write the date and place of installation of the bait trap.

6. After four days, remove the vial and bring it to the laboratory.

7. Carefully check the condition of the baited wax moth larvae. Look for the nematode infected cadavers which are either turned brick red or dark grey in colour. Generally, *Heterorhabditis* infected cadavers turn brick red in colour, whereas, *Steinernema* infected cadavers turn dark grey in colour. Generally, EPN infected cadavers do not smell, not mummify or become sloppy.

8. Separate the cadavers based on the colour of infection, rinse them in distilled water to remove the dirt on the body and transfer them to white trap.

9. For the preparation of white trap, take a watch glass (52 mm) and place with its convex side up in a large Petri dish (110x25 mm). Place a filter paper (Whatman#1, dia. 90mm) on the watch glass in such a way that, the lower edge of the filter paper should touch the bottom of the Petri dish.

10. Pour the distilled water over the filter paper and allow spreading on the Petri dish. The level of the water should be maintained such that, it should always be in contact with the edges of the filter paper only.

11. Place wax moth cadavers on the top of the watch glass. Cover the watch glass

with top cover of the Petri dish and care should be taken that; sufficient gap should be maintained between the cadavers and the top cover.

12. Place the white trap in the incubator maintained at 25 ± 10^0C and observe daily for nematode activity inside the body of the larvae. Nematodes (especially infective dauers) after exhausting the host will start emerging out of the cadavers and move to the bottom of the Petri dish containing water.

13. Decant water along with nematodes daily into a beaker and replace with fresh distilled water.

14. Clean the harvested nematodes repeatedly by washing with distilled water and store with 0.1 percent formalin solution in a conical flask. Such harvested nematodes should be further infected to fresh culture of wax moth to confirm the authenticity of EPN (Koch's postulate) infection and to obtain pure strain.

Method 2: Isolation of EPN from soil

1. Collect about a kilogram of moist soil from the identified ecosystem from up to a depth of 10-15 cm. Transfer 250 g of soil to plastic container.

2. Place 4-5 grown up wax moth larvae in each plastic container and close the lid having perforation for aeration. Leave it undisturbed for 4-5 days.

3. Open the plastic container and look for the nematode infected larvae which are either turned brick red or dark grey in colour. Generally, *Heterorhabditis* infected cadavers turn brick red in colour, whereas, *Steinernema* infected cadavers turn dark grey in colour. Generally, EPN infected cadaver do not smell, not mummified or sloppy.

4. Follow step 8 to step 14 as described above in method 1 to isolate EPNs.

Method 3: Isolation of EPN from naturally infected insect host

This type of isolation is a chance factor and need careful field observations at regular interval.

1. Look out for the dead insect larvae in the field carefully. They could be cutworms, mole crickets, cockroaches, root grubs or any other soil dwelling stages of insects such as pupae.

2. Observe the dead insects carefully to confirm the death due to EPN. This again can be done based on the colour of the dead cadaver.

3. Follow step 8 to step 14 as described above in method 1 to isolate EPNs.

Culturing of EPN

Rearing technique of laboratory host, *Galleria mellonella*

Culture of greater wax moth, *Galleria mellonella* can be maintained either on its natural diet (bee wax) or artificial food. Culturing on bee wax has certain limitations such as restricted availability of the wax and risk of infection. However, if plenty of bee wax is available in the form of abandoned bee colony, it would be economical rear on it. Otherwise, wax moths can be reared on the synthetic diet also. Advantage of this is, we can have the culture throughout the year and can multiply in large quantity.

The dietary composition of artificial diet

1. Corn meal - 400 g
2. Wheat flour - 200 g
3. Wheat bran - 200 g
4. Milk powder - 200 g
5. Yeast tablets - 100 g
6. Honey - 400 ml
7. Glycerin - 300 ml

Solid and liquid contents should be mixed separately. Later both the contents should be mixed together with constant stirring till it is homogenized.

Mass production, formulation, storage and application techniques of Entomopathogenic Nematodes

1. *In vivo* culture method

The approach consists of inoculation, harvest, concentration, and decontamination. Insect host, *G. mellonella* are inoculated on a Petri dish. After approximately 2-5 days, infected insects are transferred to the White traps. Incubate such traps in B.O.D at 25^{0}C. The nematodes starts emerging and gets collected at the bottom of the container after 10 days. Harvest the nematodes daily thereafter, repeatedly clean them with distilled water and concentrate to a required density and store.

2. *In vitro* mass production on solid medium

EPNs can also be mass produced under *in vitro* conditions on four artificial media such as modified Wout's medium, modified wheat flour medium, modified egg yolk medium and modified dog biscuit medium. Initially, the symbiotic bacterium should be isolated in a separate media.

Isolation of bacterium

Infect grownup, *Galleria* larvae with surface sterilized IJs (with 0.1% Hyamine) @ 100 per larva. Five days after the inoculation, dead hosts should be surface sterilized by rinsing in sterile water and place on an inverted glass plate kept in inoculation chamber. Dissect the larva aseptically without rupturing the midgut. Take a loopful of haemolymph and streak on NBTA plates. After streaking the haemolymph on several NBTA plates, incubate the plates at 28^{0}C in B.O.D. incubator. After two days of incubation, the bacteria, *Photorhabdus luminescens* colonies of convex, circular and mucoid with slightly irregular margins appears. This will serve as the source of inoculums for the artificial media.

Preparation of culture flasks

The conical flasks (250 ml capacity) should be washed with 0.1 per cent formalin solution and keep for sterilization at 60^{0}C for two days. The polyurethane sponge, which serves as carrier material should be cut into small pieces of 1x1 cm size and washed with four percent formaldehyde, then rinsed for 2-3 times with distilled water and dried under room temperature. Sponge pieces of 1.5 g should be used for each flask.

Preparation of artificial media

The artificial media should be prepared based on the following compositions.

Modified Wout's medium

Ingredients	Quantity (g/ml)
Nutrient broth	22.00
Yeast extract	8.00
Soy flour	360.00
Groundnut oil	260.00
Distilled water (ml)	1500.00

Modified wheat flour medium

Ingredients	Quantity (g/ml)
Wheat flour	375.00
Soy flour	125.00
Beef extract	125.00
Yeast extract	25.00
Groundnut oil	25.00
Distilled water (ml)	1500.00

Modified egg yolk medium

Ingredients	Quantity (g/ml)
Egg yolk	175.00
Soy flour	500.00
Yeast extract	50.00
NaCl	20.00
Groundnut oil	375.00
Distilled water (ml)	1500.00

Modified dog biscuit medium

Ingredients	Quantity (g/ml)
Dog biscuit	300.00
Peptone	7.50
Yeast extract	15.00
Beef extract	75.00
Groundnut oil	105.00
Distilled water (ml)	1500.00

The components of each medium should be mixed homogeneously and smeared on sponge pieces (1x1cm) at the ratio of 1.5 g: 55 g (sponge: medium) separately for each flask. Sponge pieces coated with medium should then be transferred carefully to the conical flask. Mouth of the conical flask should be cleaned and plugged with non-absorbent cotton and autoclaved at 121°C, 15 lbs for 15 min.

Inoculation with bacteria

The cultured flasks containing medium coated sponge pieces should be inoculated with bacterial suspension from nutrient broth. Inoculate each flask with 0.5 ml of bacterial suspension aseptically. After inoculation, culture flasks should be incubated at 28°C for two days to allow the multiplication and spread of the bacterial colonies over the sponge media.

Inoculation with nematodes and harvesting

The freshly harvested infective juveniles (IJs) of nematodes should be surface sterilized in 0.1 percent hyamine to prevent the contamination. These nematodes should then be inoculated at the rate of 2,000 IJs per flask, in aseptic conditions. Incubate the inoculated conical flasks at 28°C in B.O.D. incubator for 20 days. As the nutrients depletes, IJs will start emerging, which could be observed by their movement on the walls of the conical flask. The sponge pieces from such conical flasks should then be transferred to modified White trap. IJs collected in water at the base should be harvested daily in 0.1 percent formalin.

In vitro mass production in liquid medium

Nematodes can be mass produced in liquid culture using small fermentors to large bioreactors. Generally, in liquid culture, symbiotic bacteria are first introduced followed by the nematodes. Various ingredients for liquid culture media have been reported including soy flour, yeast extract, canola oil, corn oil, thistle oil, egg yolk, casein peptone, milk powder, liver extract and cholesterol. Culture time vary depending on media and species, and may be as long as three weeks though many species can reach maximum IJ production in two weeks or less. Once the culture is completed, nematodes can be harvested from media via centrifugation.

Formulation and storage

EPN can be stored in any form provided optimum moisture, temperature and oxygen is provided. However, before storage they should be cleaned off from the foreign materials like dead nematodes, dust/dirt etc.

It can be stored in the following formulations.

1.   Aqueous suspension

Freshly emerged IJs collected from White traps should be ascertained for 100 percent activity. Allow them to settle at the bottom of the container and decant the supernatant. Adjust the density to 10,000 IJs per ml. Nematode suspension of 25 ml should be pipetted out in 250 ml conical flask and closed with cotton plug. The Petri dish containing IJs should be aerated using aquarium pump thrice in a week and also change the water every week. The mortality or the activity of the nematodes can be checked by pipetting out 1 ml of the aqueous suspension into a counting chamber or by serial dilution method.

2.   Talc based formulation

Talc based formulation can be prepared by using autoclaved talc to which nematode suspension of 10,000 IJs/ml should be added to the mixture @ 4 ml/25 g of talc and mixed well to distribute the nematodes evenly in the medium. 25 g of formulated talc can be stored in plastic press bags (7x2.5cm). Viability of nematodes in the talc formulation under storage can be verified by dispersing 0.5g of talc in 5ml of water and observing under dissection microscope. Straight and immovable nematodes should be considered as dead whereas, moving individuals irrespective of activity should be considered as live.

3. Sponge formulation

Polyether–polyurethane sponge, used for storing IJs, should be cut into small pieces of 2.5x1.5 cm size and washed with four percent formaldehyde. Then rinsed 2 to 3 times with distilled water and dried under room temperature. Further, nematodes should be injected on such clean sponges and packed in polythene bags to prevent desiccation. Viability of the infective juveniles should be verified periodically by hand squeezing the sponge in a small volume of water from the sponge.

4. Encapsulation with sodium alginate gel formulation

The IJs can be encapsulated in sodium alginate. Solution of the gel matrix should be prepared by dissolving 2g of sodium alginate in 100ml water with thorough blending for 10-15min. To the gel matrix, different concentration of nematodes can be added and mixed thoroughly. Drops of this solution through 1ml hypodermic syringe should be placed into the complexing solution (100mM solution of Calcium chloride. Continuously stir the solution to get discrete capsules of calcium alginate of 0.4cm diameter. Capsules are allowed to harden for 20-30 min and then should be separated from the complexing solution by sieving. Rinse the capsules in distilled water and store in plastic Petri plates (90 diax10 h mm) lined with wet blotting paper.

Application techniques

The EPN can be applied with nearly all agronomic or horticultural ground equipment including pressurized sprayers, mist blowers and electrostatic sprayers or as aerial sprays.

» It is important to ensure adequate agitation during application

» For small plot applications, hand-held equipment (e.g., water cans) or back-pack sprayers may be appropriate

» Various formulations for EPN may be used for applying. EPNs in aqueous suspension including activated charcoal, alginate and polyacrylamide gels, clay, peat, polyurethane sponge, vermiculite and water dispersible granules (WDG) are applied

» Also to be effective, EPN usually must be applied to soil at minimum rates of 2.5x109 IJs/ha (=25/cm^2) or higher

»   Generally, as long as environmental conditions are conducive, nematode
    populations will remain high enough to provide effective pest control for 2 to
    8 weeks after application

22

Mass Production and Formulation of Entomopathogens

Rajeshwari R. and Vikram Appanna

University of Horticultural Sciences, Bagalkot

Introduction

A group of fungi that kill an insect by attacking and infecting its insect host is called entomopathogenic fungi. The main route of entrance of the entomopathogen is through integument and it may also infect the insect by ingestion method or through the wounds or trachea. The divisions of fungi are Ascomycota, Zygomycota and Deuteromycota, and the divisions Oomycota and Chytridiomycota were also included in the previous classification of fungi. More than 700 species of fungi from around 90 genera are pathogenic to insects.

The most important insect infecting species occur in *Aspergillus, Beauveria, Metarhizium, Hirsutella, Aschersonia, Culicinomyces, Lecanicillium, Paccilomyces, Tolypocladium* and *Sorosporella*.

Fungi represent a diverse group of insect pathogens. The insects attacked by the fungus die shortly after the fungus begins to develop in the haemocoel. The rhinoceros beetle, *Oryctes rhinoceros* is one of the serious and important pests of coconut and has a wide distribution and persistent occurrence in all the coconut growing areas in the country. The adult beetle cause severe damage to the growing palms by feeding on the tender fronds and crowns and resulting in stunting of the trees.

The damage to the spathe results in the loss of nuts. Young seedlings are sometimes killed outright. Since the insect breeds in the farmyard manure and fallen coconut trees, the control measures have to be directed at the breeding sites as well as on the trees. The chemical control measures adopted against this pest are always costly, tedious and have to be repeated. Hence an easy method utilizing the safe and specific fungus, *Metarhizium anisopliae* for the management of this pest is aimed at.

1. Mass Production of *Metarhizium anisopliae*

» The colony of *M. anisopliae* appears white when young, but as the conidia mature, the colour turns to dark green.

» The conidiophores are branched, and the initial conidium is produced at the distal end of the conidiophores.

» A chain of conidia is formed on each conidiophore with the youngest conidium being adjacent to the conidiophore.

» The mass of spore chains becomes so dense and coheres with each other to produce prismatic masses of columns of spore chains.

Mass production in molasses yeast broth media

» Molasses yeast broth should be prepared by mixing 30 g of molasses and 5g of yeast extract in 1000ml of water and the pH is adjusted to 6.0. It is transferred to 2 litre flasks.

» The conical flasks are plugged with cotton and autoclaved for 20 min at 15 psi.

» The flasks are allowed to cool and taken to laminar flow chamber for inoculation.

» From a clean uncontaminated mother culture in slant loopful quantities of *M. anisoliae* spores are transferred aseptically.

» The flasks are incubated at room temperature. The spores can be harvested in a fortnight.

Mass production in solid substrate media

1. Fill polypropylene bags (25x35 cm) with 500g of broken rice.

2. Soak it in water overnight and then decant water in the morning.

3. Sterilize the soaked grains for 30 min in an autoclave or pressure cooker.

4. *Metarhizium* mother culture (2 days old *M. anisopliae*) culture grown on potato dextrose agar) should be used for inoculation.

5. Inoculate the sterilized grains with *Metarhozium* spore suspension (2×10^8 conidia/ml) grown in potato dextrose broth (5 ml), plug the bags, mix the inoculated culture bag gently and incubate for 25 days.

6. *Metarhizium* produces dark green spores and coats on the grains. These grains are used for preparation of formulation.

Preparation of talc formulation

1. PP bags showing full sporulation arc cut open and the grains arc crushed to fine powder and mixed with sterilized talc in proportion of 1:2 or 1:3 and dried under aseptic conditions for 24 hrs.

2. The spore load in the formulation may be 106-1010 spore/g.

3. The formulation can be diluted with water to get spore concentration of 1010 spore/liter for foliar sprays. Tween-80 (@ 0.01%) should be added to the spore suspension for uniform spread of spores on plant surface.

2. Mass production of *Beauvaria Bassiana*

The fungus is otherwise called as white muscardine fungus. The fungus spores and mycelia are milky white and found sprouting on the body of lepidopterous insects like *Helicoverpa armigera*, *Spodoptera litura* and *Anadevidia peponis*.

Fermentation strategies

Effective control with fungal entomopathogens using inundation biocontrol requires an understanding of the ecology of the target insect, fungal pathogen, and the insect-pathogen interaction. Majority of the commercialized products are comprised of conidial preparations of *B. bassiana*, presumably using solid substrate fermentation production processes. *B. bassiana* have a very broad insect host range with many isolates producing high concentrations of aerial conidia when grown on nutrient rich, solid substrates (Pereira and Roberts, 1990). The use of conidia as fungal product is warranted as they are a naturally-infective propagule. There are ecological and environmental conditions in which the use of conidia may not be the best choice for insect biocontrol in agricultural or urban settings.

Isolation of *B. bassiana*

From infected insect: The fungus can be isolated from the infected insects in two ways, either from the hemolymph or directly from the fungal growth on the surface.

Isolation from hemolymph: Infected insects are first surface sterilized by immersing them in 5% hypochloride solution for a few minutes and rinsed in sterile water three times. Then specimen is cut open in a sterile plate and small portion of infected hemolymph/ tissue is streaked on PDA (Potato, 200.0g, Dextrose 10.0g, Agar 15.0g, and Distilled water, 1000ml) for isolation of *B. bassiana*. The pure colonies of *B. bassiana* may be sub-cultured on PDA/SMAY slants and stored in a refrigerator.

Direct isolation from surface of the infected insects: Mycelium and spores can be removed from a fresh specimen and placed directly on the PDA/SMAY/SDAY medium. However, this isolate should he compared with that obtained from infected hemolymph/tissue.

Growth on solid substrates in polypropelene bags: *B. bassiana* can be grown easily on cereal grains like Sorghum, Rice and Maize etc.

1. The grains are soaked in water over night and next day the water is decanted and crushed slightly in a blender for a few seconds.

2. 200 g of crushed grains are taken in PP bags of (30X20cm) and sterilized at 121oC temperature for 30 minutes.

3. After cooling the bags are cut open and inoculated with 3 ml of suspension (106 spores/ml) prepared from fresh culture and the cut portion is plugged with sterile cotton wool.

4. The inoculated grains are spread out as thin layer in the bags and incubated 7-10 days at room temperature or below 30°C.

Preparation of talc formulation

1. PP bags showing lull sporulation arc cut open and the grains arc crushed to fine powder and mixed with sterilized talc in proportion of 1:2 or 1:3 and dried under aseptic conditions for 24 hrs.

2. The spore load in the formulation may be 10^6-10^{10} spore/g.

3. The formulation can be diluted with water to get spore concentration of 1010 spore/liter for foliar sprays. Tween-80 @0.01% should be added to the spore suspension for uniform spread of spores on plant surface.

3. Mass production of *Verticillium lecanii*

Vertilcillium lecani commonly called as white halo fungus is found sprouting on the body of coffee green scale *Coccus viridis*. The fungus is known to cause epizootic when the environmental conditions are favourable.

Production procedure

» The fungus is multiplied on cheap media for large scale production.

» Sorghum grains (40 g) is washed in water and transferred to 250ml conical flask and 15 ml of distilled water is added.

» The conical flasks are plugged with cotton and autoclaved for 20 min at 15 psi.

» The flasks are allowed to cool and taken to laminar flow chamber for inoculation.

» From a clean uncontaminated mother culture in slant loopful quantities of *V. lecani* spores are transferred aseptically.

» The flasks are incubated at room temperature. The spores are obtained in a fortnight.

Preparation of talc formulation
1. PP bags showing full sporulation arc cut open and the grains arc crushed to fine powder and mixed with sterilized talc in proportion of 1:2 or 1:3 and dried under aseptic conditions for 24 hrs.

2. The spore load in the formulation may be 106-1010 spore/g.

3. The formulation can be diluted with water to get spore concentration of 1010 spore/liter for foliar sprays. Tween-80 (@ 0.01%) should be added to the spore suspension for uniform spread of spores on plant surface.

4. Mass production of *Nomuraea rileyi*

Nomuraea rileyi is an entomopathogenic fngus of cosmopolitan occurrence, primarily infecting Lepidoptera and particularly the economically important, polyphagous noctuid pests. Preliminary identification of *N. rileyi* is possible by looking for malachite-green colouration on the insect surface. When viewed under the microscope, conidiophores are seen bearing dense whorls of phialides *i.e.*, conidiogenous cells that are short necked. Conidia of *N. rileyi* are broadly ellipsoidal to cylindrical with a size of 3.5-4.5 x 2.0-3.1 μm. *N. rileyi* is a fastidious fungus and can be best multiplied on Saboraud's dextrose agar or maltose agar fortified with yeast extract.

Solid-state fermentation is very common method for mass production of fungal bio-agents under laboratory conditions. Various agricultural wastes and by-products could be used for mass production eg. grain bran, wheat straw, wheat bran, wheat bran-saw dust, wheat bran-peat, sorghum grain etc. *N. rileyi* can be cultivated on different media - sorghum, barley, wheat bran etc. Aerial conidia produced are indistinguishable in morphology and infectivity from those produced on the surface of insect cadavers.

On Sorghum

» Crush the sorghum grains in a mixie to get broken pieces. Weigh 25 g of the crushed sorghum in a 250 ml conical flask and add 22.5 ml of 1% yeast extract solution. Soak overnight at 25°C. Plug the flasks and autoclave. Immediately after cooling, break the clump of sorghum aseptically using a blunt forceps. Add dry spore of *Nomuraea rileyi* to the sorghum using a micro-spatula. Shake well to disperse the spore evenly in the medium and incubate the flasks at 25°C in the dark. Mycelial growth is initiated on 4th or 5th day and continues for 3-4 days. Sporulation is initiated at 7-8 days after inoculation and continues for 3-4 days. The flasks are then transferred to refrigerator and stored for 4-5 days (sporulation continues in the refrigerator). The substrate along with the fungus is the shade dried and sieved through a muslin cloth to obtain pure conidia of

N. rileyi. Scale-up production can be undertaken in polypropylene bags plugged with non-absorbent cotton.

On Barley

» 5 g of crushed barley is taken in a 250 ml conical flask and 45 ml of distilled water containing 0.125 g of yeast extract. The flask is plugged with non-absorbent cotton and autoclaved. After cooling, the flasks are inoculated aseptically with dry spore of *N. rileyi*, mixed well using a sterile spatula. The flask is plugged and incubated at 25⁰C in the dark till mycelial growth and sporulation is complete.

» For scale-up production, the medium can be poured into glass trays, transferred to a polythene cover plugged with non-absorbent cotton at the open end and sterilized. After cooling the medium can be inoculated with *N. rileyi* conidia aseptically, mixed well with a sterile spatula. The open end is again plugged with the cotton and the bag is incubated i at 25⁰C in the dark till mycelial growth and sporulation is complete.

On polished rice grains

» Put 300 g boiled rice (for 10 minutes) into a bag with a synthetic sponge in its opening or a mixture of raw rice and water in 1:2 ratio, sterilize and cool. Cut the other end, inoculate dry spore aseptically and seal. Incubate at 25⁰C. Shade dry the substrate and sieve to get the conidial powder.

On insects

» *Nomuraea rileyi* can also be multiplied without much difficulty on 7-8 days old *S. litura* larvae on castor leaves treated with *N. rileyi* conidia. Larval death results in 6-7 days after exposure. Mycelial growth on mummified cadavers occurs within 24 h followed by sporulation 1-3 days later. Each larva yields about $2 - 4 \times 10^{10}$ conidia. Sporulating cadavers can be dispersed in the field to increase the inoculum before the peak incidence of the pest. The feasibility of mass multiplication on insects needs to be explored.

Preparation of talc formulation

After 25 days of inoculation, the fungal growth in the broth was collected along with the medium and blended in the mixer for 1-2 min to get a homogenous slurry. Then it is strained through muslin cloth to remove debris under aseptic conditions. This contained 2x10^9 conidia/ml. In case of grain substrate, the spore mass along with grain carrier was taken out from bags after 25 days and air dried under laminar flow for three days. The spore count per gram should be 1.8x10^9 conidia. Tal powder was sterilized in autoclave. After cooling, the fungal slurry of known spore strength was mixed in the talc powder to obtain the formulation of desired strength. Then this mixture was dried under shade/ laminar hood for 3-4 days under aseptic conditions, sieved through 18 mesh screen and packed in sterile polypropylene bags. Similarly grain spore powder mixed in sterilized talc powder carrier and packed. These formulations can be used in the field.

5. Mass production of the Nuclear Polyhedrosis Virus (NPV) of *Helicoverpa armigera* and *Spodoptera litura*

In India, *H. armigera and S. litura are* of major importance damaging a wide variety of food, fibre, oilseed, fodder and horticultural crops. The nuclear polyhedrosis virus of *H. armigera* (HearNPV) and *S. litura* (SpliNPV) are currently used for the management of *H. armigera* and *S. litura* respectively on different crops. Mass production of Nuclear Polyhedrosis Virus (NPV) on commercial scale is restricted to *in vivo* procedures in host larvae which are obtained by

» Field collection of *H. armigera and S. litura*

» Mass culturing in the laboratory in semisynthetic diet

Rearing of laboratory hosts

Rearing of larvae in the natural host plant will involve frequent change of food at least once a day during the incubation period of 5-9 days increasing the handling time. In order to reduce the cost, field collected larvae are released into semi synthetic diet treated with virus inoculum. Mass culturing of insects in semi synthetic diet involves high level of expertise, hygiene and cleanliness. Collection of a large number of larvae in optimum stage (late IV / early V instars) is time-consuming and can be expensive in terms of labour and transportation costs.

Production procedure

The NPV of *H. armigera* and *S. litura* are propagated in early fifth instar larvae of *H. armigera* and *S. litura* respectively. The virus is multiplied in a facility away from the host culture laboratory. The dose of the inoculum used is 5 x 105 polyhedral occlusion bodies (POB) in 10 ml suspension. The virus is applied on to the semisynthetic diet (lacking formaldehyde) dispensed previously in 5 ml glass vials. A blunt end polished glass rod (6 mm) is used to distribute the suspension containing the virus uniformly over the diet surface. Early fifth instar stage of larvae are released singly into the glass vials after inoculation and plugged with cotton and incubated at a constant temperature of 25oC in a laboratory incubator. When the larvae exhausted the feed, fresh untreated diet is provided. The larvae are observed for the development of virosis and the cadavers collected carefully from individual bottles starting from fifth day. Approximately, 200 cadavers are collected per sterile cheese cup (300 ml) and the contents are frozen immediately. Depending upon need, cadavers are removed from the refrigerator and thawed very rapidly by agitation in water.

Processing of NPV

The method of processing of NPV requires greater care to avoid losses during processing. The cadavers are brought to normal room temperature by repeatedly thawing the container with cadaver under running tap water. The cadavers are homogenized in sterile ice cold distilled water at the ratio 1: 2.5 (w/v) in a blender or precooled all glass pestle and mortar. The homogenate is filtered through double layered muslin and repeatedly washed with distilled water. The ratio of water to be used for this purpose is 1: 7.5-12.5 (w/v) for the original weight of the cadaver processed. The left over mat on the muslin is discarded and the filtrate can be semi-purified by differential centrifugation. The filtrate is centrifuged for 30-60 sec. at 500 rpm to remove debris. The supernatant is next centrifuged for 20 min at 5,000 rpm. Then the pellet containing the polyhedral occlusion bodies (POB) is suspended in sterile distilled water and washed three times by centrifuging the pellet in distilled water at low rpm followed by centrifugation at high rpm. The pellet finally collected is suspended in distilled water and made up to a known volume, which is necessary to calculate the strength of the POB in the purified suspension.

6. Mass production and formulation of *Bacillus thuringiensis*

Bacillus thuringiensis (Bt) is the most successful microbial pesticide ever registered. It is a gram-positive bacterium forming elliptical spores, contained in unswollen sporangia, and a parasporal body (crystal) which appears mainly as a bipyramidal shape. Bt is a complex species divisible into subspecies and H-serotypes by serological and biochemical tests. Bt isolates/strains produce several insecticidal toxins, two of which are used in agriculture. The relative activity of each isolate against different insect species "spectrum activity" arrives partly from the combined effects of the potencies of the varying concentrations of the different insecticidal toxins that it produces. The δ-endotoxin of different isolates of Bt can kill different insect species or differ in the degree of their activity toward them. Maximum toxin production can be achieved only by careful attention to the interaction of fermentation conditions, media and the isolates involved - there is, for example, no one medium best suited to all isolates.

Culturing on media

Bt is easily multiplied on Luria-Bertani agar (LB), Nutrient agar and T3 agar media. Pure culture of Bt is maintained on these agar slopes while it is multiplied in the broth (medium without agar) of these media in flasks for testing against insects. After inoculation, the flasks plugged are with sterile cotton and placed on a shaker at 30⁰C. Time taken for complete growth, sporulation and lysis is 48-72 h depending on the strain/isolate. The Bt spores and crystals are then recovered from the broth through centrifugation, the resultant pellet is dried and milled to get a powder, which is used for bioassays.

LB agar Medium (Luria-Bertani Medium) (pH 7.0-7.2):

Tryptone-10.0 g, Yeast extract-5.0 g, Sodium chloride-5.0 g, Agar-15.0 g, Water-1000 ml

T3 agar medium (pH 6.9)

Tryptone-3.0 g, Tryptose-2.0 g, Yeast extract-1.0 g, Magnesium chloride-0.005 g, Sodium dihydrogenphosphate-6.9 g, Di-sodium phosphate-8.9 g, Agar-20.0 g, Water-1000 ml

Nutrient agar (pH 7.2)

Peptone-10.0 g, Sodium chloride-5.0 g, Yeast extract-5.0 g, Agar-15.0 g, Water-1000 ml.

Growth phases

After inoculation there is an initial lag phase followed by the exponential phase during which intensive growth of the culture takes place. This phase persists up to the 16th-18th hour. At the end of this exponential phase, sporulation is initiated and the cells enter into the stationary phase. Spores start appearing within the cells, together with parasporal inclusions of crystalline toxin. Sporulation is complete by 48-50 hours. This is followed by the lytic/death phase wherein the cells are subject to lysis, liberating spores and crystalline inclusions into the fermentation liquid. Some 90-98% of all spores and inclusions are liberated after 60-72 hours. The changes of pH during fermentation depend on the composition of the medium used. After sterilization of the fermentation medium the pH should ideally be 6.8-7.2. After inoculation, during the exponential phase, acids being formed from the saccharides, pH will drop to 5.8-6.0 (after 10-12 hours), will rise again to about 7.5 (after 25 hours), further rise to 8.0 (after 30-35 hours hours) and ultimately to 8.8 (50-60 hours). pH during the fermentation period is maintained around 7.5 by the addition of 1N sodium hydroxide/1N hydrochloric acid.

Fermentation Techniques for Mass Production

After the growth and sporulation processes have been thoroughly investigated and tested at laboratory level, mass production of Bt can be undertaken on an industrial scale with various raw materials. The standard method of production of microorganisms is the process of fermentation. There are many types of fermentation; the two most common are "submerged" and "solid". For optimal growth of Bt important factors are pH, temperature and oxygen. It is possible to control these parameters in fermentors. Hence, Bt is traditionally produced in liquid media in large capacity industrial fermentors employing the method of submerged or deep tank fermentation, which is as the name implies growth of microorganism in a fully liquid system.

Submerged or deep-tank fermentation is, as the name implies, a growth of micro-organisms in a fully liquid system. There are a number of advantages to fully liquid systems which include the ability to hold temperature and pH constant, the ability to pump large quantities of air into the system and disperse it by means of stirring impellers, and the ability to generate reasonably homogeneous conditions to maximize the growth of micro-organism.

In the beginning of small scale processing the fermenters used are the seed fermenters of 20-40 litres capacity which contain 10-25 litres of nutrient medium. Culture from the shaker or from laboratory fermenter (1-3 litres), to the amount of 1-3% of volume of the medium, serves for inoculation. The seed tanks can be made of glass or stainless steel; the fermentation tanks are made of stainless steel. An aeration ring is used to aerate the culture, together with an agitator or an air outlet under a propeller. The air volume used for aeration should correspond to between 1/2 and full volume of the cultivation medium.

Unless foam suppressing agents are used, the fermentation liquid will be subject to intensive foaming at the beginning of fermentation and also at the sporulation time (after about 24 hours). Silicone defoaming agents can be used to mitigate foam formation.

Solid-state fermentations (SSF) are relatively easy to develop on a small scale. Scaling them up to the sizes necessary for commercial production presents numerous problems, aeration becomes a major difficulty as the volume of the solid mass increases more rapidly than the available surface area. Substrates like wheat bran and cotton seed meal have been employed successfully for multiplication Bt through SSF.

Low-cost mass production of *Bacillus thuringiensis* var. *kurstaki* isolate on the principle of solid-state fermentation can be successfully undertaken on a solid medium containing wheat bran (150 g), molasses (3.6 g), yeast extract (0.72g), potassium salts (0.36 g of di-potassium hydrogen orthophosphate and potassium di-hydrogen phosphate) and 250 ml distilled water. The medium (pH 7.2-7.5) is autoclaved in polythene covers and dispensed aseptically into sterile plastic tubs after cooling. Seed culture of the isolate (multiplied in nutrient broth) is added to the medium, mixed thoroughly and closed air tight with a polythene sheet. The tubs are incubated for a period of 60-65 hours at 30⁰C with intermittent aerations in a laminar airflow. After complete lysis, sterile distilled water is added to the fermented medium and filtered through a medium muslin cloth. The filtrate is centrifuged to remove the water-soluble β-exotoxin, which remains in the supernatant that is discarded. The pellet is shade dried, powdered and passed through a sieve to get the technical material. The filtration residue is shade dried, powdered and passed through a sieve to get the bran-based carrier.